Prentice Hall Interactive MATH

Key Concept Activity Lab Workbook

Martha Haehl
Maple Woods Community College

James M. Sullivan
Massachusetts Bay Community College

Basic Math

K. Elayn Martin-Gay
University of New Orleans

PRENTICE HALL, Upper Saddle River, NJ 07458

Editor in Chief: ***Jerome Grant***
Director of Product Development: ***Christine B. Hoag***
Editor in Chief, Development: ***Carol Trueheart***
Development Editor: ***Elaine Page***
Associate Editor, Math/Statistics Media: ***Audra Walsh***
Producer: ***Sara Wilson***
Special Projects Manager: ***Barbara A. Murray***
Production Editor: ***Barbara A. Till***
Design Director: ***David Roh***
Production Coordinator: ***Alan Fischer***
Supplement Cover Manager: ***Paul Gourhan***
Supplement Cover Designer: ***Liz Nemeth***

Upper Saddle River, NJ 07458

Printed in the United States of America

10 9 8 7 6 5 4 3 2 1

ISBN 0-13-086410-2

Prentice-Hall International (UK) Limited, *London*
Prentice-Hall of Australia Pty. Limited, *Sydney*
Prentice-Hall Canada, Inc., *Toronto*
Prentice-Hall Hispanoamericana, S.A., *Mexico*
Prentice-Hall of India Private Limited, *New Delhi*
Prentice-Hall (Singapore) Pte. Ltd.
Prentice-Hall of Japan, Inc., *Tokyo*
Editora Prentice-Hall do Brasil, Ltda., *Rio de Janeiro*

Table of Contents

Interactive Math Objective Chart—Linking the Workbook Exercises to Objectives

Each exercise in this Key Concept Activity Lab Workbook is supported by Objectives within Interactive Math—Basic Math. These Objectives and topics appear below.

Key Concept Activity	Topics and Objectives Covered			
	Extension Exercise	**Open Environment Activity**	**Conceptual Exercise**	**Group Activity**
More Bash for the Buck (Chapter 1) pp. 1-10	1.1.D, 1.5.A, 1.5.B, 1.5.D, 1.6.A, 1.6.B, 1.6.C reading tables, multiplying and dividing whole numbers, solving problems, long division, properties of multiplication	1.1.A, 1.1.B, 1.1.D, 1.2.A, 1.2.C, 1.3.A, 1.3.B, 1.3.C, 1.4.A, 1.4.B, 1.4.C, 1.6.D, 1.7.A, 1.7.B place value, standard form, reading tables, adding and subtracting whole numbers, solving problems, estimating, borrowing, rounding, averages, solving problems with more than one operation	1.5.B, 1.5.D, 1.6.A, 1.6.B, 1.6.C, 1.7.A multiplying and dividing whole numbers, long division, problem solving	1.1.A, 1.1.B, 1.1.D, 1.2.A, 1.2.C, 1.3.A, 1.3.B, 1.3.C, 1.4.A, 1.4.B, 1.4.C, 1.7.A, 1.7.B place value, standard form, reading tables, adding, subtracting, and dividing whole numbers, rounding and estimating, solving problems with more than one operation, borrowing, problem solving
Cookie Cutting (Chapter 2) pp. 11-20	1.1.D, 1.2.A, 1.2.C, 1.5.B, 1.5.C, 1.5.D, 2.1.A, 2.3.A, 2.4.B, 2.4.C adding and multiplying whole numbers, reading tables, solving problems, area of a rectangle, fractions, simplest form, multiplying fractions with whole numbers	1.1.D, 2.1.A, 2.2.A, 2.3.A, 2.3.C, 2.4.A, 2.4.B, 2.4.C tables, spreadsheets, fractions, factoring numbers, simplest form, solve problems, multiplying fractions, multiplying fractions and mixed or whole numbers	1.6.A, 1.6.C, 2.1.A, 2.3.A, 2.3.C, 2.4.A, 2.4.C dividing whole numbers, problem solving, fractions, simplest form, multiplying fractions	2.1.A, 2.1.C, 2.1.D, 2.1.E, 2.3.A, 2.3.C, 2.4.A, 2.4.B, 2.4.C Fractions, improper fractions to mixed numbers and vice versa, simplest form, problem solving, multiplying fractions, multiplying fractions with mixed or whole numbers
Playing the Market (Chapter 3) pp. 21-30	1.1.D, 3.2.A, 3.2.B, 3.2.C, 3.4.A, 3.4.B, 3.4.C reading tables, adding and subtracting mixed numbers, equivalent fractions, LCM, problem solving	1.1.D, 2.3.A, 2.4.A, 2.4.B, 2.4.C, 3.3.A, 3.3.C, 3.4.A, 3.4.C tables, spreadsheets, adding and multiplying unlike fractions, adding and multiplying mixed numbers, simplest form, problem solving	2.1.B, 2.3.B, 3.1.A, 3.1.B, 3.1.C, 3.2.A, 3.3.A, 3.3.B, 3.3.C, 3.4.A, 3.4.C fractions and shaded figures, equivalent fractions, LCD, adding and subtracting unlike and like fractions, problem solving, adding mixed numbers	2.1.A, 2.1.B, 2.1.D, 2.1.E, 2.4.A, 2.4.B, 2.4.C, 2.5.A, 2.5.B, 2.5.C, 2.5.D, 3.1.A, 3.1.C, 3.3.A, 3.3.C, 3.4.A, 3.4.C fractions and shaded figures, mixed numbers to improper fractions and vice versa, multiplying and dividing fractions and mixed numbers, adding fractions, adding mixed numbers

Key Concept Activity	Topics and Objectives Covered			
	Extension Exercise	Open Environment Activity	Conceptual Exercise	Group Activity
Going the Distance (Chapter 4) pp. 31-40	1.1.D, 2.4.B, 2.4.C, 4.1.A, 4.1.B, 4.1.C, 4.1.D, 4.2.A, 4.2.B, 4.4.A, 4.4.D, 4.5.A, 4.5.C, 4.7.A, 4.7.B, 4.7.C tables, multiplying fractions and mixed or whole numbers, problem solving, decimals, decimals to fractions and vice versa, rounding decimals, multiplying and dividing decimals, comparing decimals and fractions	1.1.D, 4.1.A, 4.2.A, 4.2.B, 4.4.A, 4.4.D, 4.5.A, 4.5.C, 4.7.A, 4.7.B, 4.7.C tables, spreadsheets, decimals, rounding decimals, multiplying and dividing decimals, comparing decimals and fractions, problem solving	4.1.A, 4.2.A, 4.2.B, 4.3.A, 4.3.C, 4.4.A, 4.4.C decimals, rounding decimals, adding decimals, multiplying decimals, problem solving	4.1.A, 4.3.A, 4.3.C, 4.4.A, 4.4.D decimals, problem solving, adding and multiplying decimals
By Popular Vote (Chapter 5) pp. 41-50	1.1.D, 1.4.A, 5.1.A, 5.3.A, 5.3.B, 5.3.C, 5.4.A tables, rounding, ratios as fractions, proportions, checking proportions for truth, finding unknown values in proportions, problem solving	1.1.D, 1.4.A, 1.4.B, 1.4.C, 4.2.A, 4.2.B, 4.4.A, 4.4.D, 5.1.A, 5.1.B, 5.2.A, 5.2.B, 5.2.C, 5.3.A, 5.3.B, 5.3.C, 5.4.A tables, spreadsheets, rounding and estimating, problem solving, multiplying and rounding decimals, ratios as fractions, rates as fractions, proportions, unit rates and unit prices, finding unknowns in proportions	1.1.D, 5.1.A, 5.1.B, 5.3.A, 5.3.B, 5.3.C, 5.4.A tables, ratios, problem solving, proportions, finding unknowns in proportions	1.1.D, 4.1.A, 4.1.B, 4.1.C, 4.1.D, 4.2.B, 5.1.A, 5.1.B tables, decimals to fractions and vice versa, problem solving, rounding decimals, ratios as fractions, simplifying ratios
Free Throw (Chapter 6) pp. 51-60	1.1.D, 4.1.B, 4.1.C, 4.1.D, 4.2.B, 5.1.A, 5.1.B, 6.1.A, 6.1.C, 6.2.B, 6.2.C, 6.3.A, 6.3.B, 6.4.B, 6.5.A tables, ratios as fractions, rounding decimals, problem solving, percents, writing decimals as percents, writing fractions as percents, percents as equations, proportions	1.1.D, 5.1.A, 6.1.A, 6.2.A, 6.2.B, 6.4.B, 6.5.A, 6.5.B tables, spreadsheets, ratios as fractions, percents, writing percents as fractions and vice versa, percents and problem solving, percent increase and decrease	6.1.A, 6.2.B, 6.2.C, 6.3.B, 6.5.A, 6.5.B percent, writing fractions as percents, problem solving with percents, percent increase and decrease	1.1.D, 6.1.A, 6.2.A, 6.3.A, 6.3.B, 6.4.A, 6.4.B, 6.5.A tables, percents as fractions, percent problems as equations, percent proportions as equations

Key Concept Activity	Topics and Objectives Covered			
	Extension Exercise	Open Environment Activity	Conceptual Exercise	Group Activity
Sea Turtles (Chapter 7) pp. 61-70	2.4.A, 2.4.B, 2.4.C, 5.2.A, 5.3.A, 5.3.C, 6.1.B, 6.2.A, 6.3.A, 6.3.B, 6.5.A, 7.1.A, 7.1.B, 7.1.C, 7.1.D, 7.2.A, 7.2.B, 7.2.C, 7.2.D multiplying fractions and whole numbers, problem solving, writing percents as decimals or fractions, writing percent problems as equations, rates as fractions, proportions and unknown values, Metric to U.S. units in length, Metric to U.S. units in weight, unit fractions	1.1.D, 4.2.B, 5.2.B, 5.3.A, 5.3.C, 5.4.A, 7.1.A, 7.1.B, 7.1.C, 7.1.D tables, spreadsheets, rounding decimals, unit rates, proportions, unknowns in proportions, U.S. and metric units of length, problem solving	5.2.B, 5.3.A, 5.3.C, 5.4.A, 7.2.A, 7.2.B, 7.2.C, 7.2.D, 7.3.A, 7.3.B, 7.3.C, 7.3.D unit rates, proportions, unknowns in proportions, problem solving, U.S. and metric units of weight and capacity	5.3.A, 5.3.C, 5.4.A, 7.4.A, 7.4.B proportions, unknowns in proportions, problem solving, U.S. and metric systems of temperature
Home Run (Chapter 8) pp. 71-80	8.2.A, 8.3.A, 8.4.A plane figures, perimeter, area	1.1.D, 4.2.B, 8.2.A, 8.3.A, 8.4.A tables, spreadsheets, rounding decimals, plane figures, perimeter, area	8.1.B, 8.2.A, 8.3.A, 8.4.A, 8.6.A, 8.6.C classifying angles, plane figures, perimeter, area, square roots, Pythagorean Theorem	8.2.A, 8.2.B, 8.3.A, 8.5.A plane figures, solid figures, perimeter, volume
The Sneeze Factor (Chapter 9) pp. 81-90	9.5.A, 9.5.B counting, probability of an event, possible outcomes	1.1.D, 6.1.A, 6.2.A, 6.2.B, 6.4.B, 6.5.A, 9.2.A, 9.2.B, 9.3.A, 9.3.B, 9.4.A, 9.4.B, 9.4.C tables, spreadsheets, percents, writing percents as fractions and vice versa, percents and problem solving, circle or pie graphs, histograms, frequency distributions, mean, median, mode	4.2.B, 6.1.A, 6.2.B, 6.2.C, 6.3.B, 6.5.A, 9.5.A, 9.5.B rounding decimals, percent, writing fractions as percents, problem solving with percents, counting, probability of an event	4.2.B, 6.1.A, 6.2.B, 6.2.C, 6.3.B, 6.5.A, 9.5.A, 9.5.B rounding decimals, percent, writing fractions as percents, problem solving with percents, tree diagram, counting, probability of an event

Key Concept Activity	Topics and Objectives Covered			
	Extension Exercise	Open Environment Activity	Conceptual Exercise	Group Activity
Fire and Ice (Chapter 10) pp. 91-100	10.1.A, 10.1.C, 10.2.A, 10.2.B, 10.3.A, 10.3.B, 10.3.C, 10.4.A, 10.4.B, 10.4.C, 10.5.A problem solving, signed numbers, operations on signed numbers, order of operations	1.1.D, 4.2.B, 10.1.A, 10.1.C, 10.1.D, 10.1.E, 10.2.A, 10.2.B, 10.3.A, 10.3.B, 10.3.C tables, spreadsheets, rounding decimals, problem solving, signed numbers, absolute value, opposites, adding and subtracting signed numbers	4.2.B, 10.1.A, 10.1.C, 10.2.A, 10.2.B, 10.3.A, 10.3.B, 10.3.C rounding decimals, problem solving, signed numbers, adding and subtracting signed numbers	10.1.A, 10.1.C, 10.2.A, 10.2.B, 10.3.A, 10.3.B, 10.3.C, 10.4.A, problem solving, signed numbers, operations on signed numbers
A Daily Dose (Chapter 11) pp. 101-110	1.1.D, 11.1.A, 11.1.B, 11.1.C, 11.3.A, 11.4.A tables, variables, algebraic expressions, like terms, multiplying expressions, multiplication property	1.1.D, 4.2.B, 11.1.A, 11.1.B, 11.1.C, 11.2.A, 11.2.B, 11.3.A, 11.4.A tables, spreadsheets, rounding decimals, problem solving, variables, algebraic expressions, like terms, multiplying expressions, addition and multiplication properties	1.1.D, 4.2.B, 8.3.B, 11.1.A, 11.1.B, 11.1.C, 11.3.A, 11.4.A rounding decimals, circumference, tables, variables, algebraic expressions, like terms, multiplying expressions, multiplication property	1.1.D, 4.2.B, 11.1.A, 11.1.B, 11.1.C, 11.3.A, 11.4.A tables, rounding decimals, variables, algebraic expressions, like terms, multiplying expressions, multiplication property

Directions on Using the Open Environment (Interactive Math Toolkit) Tools

To go to the Open Environment (Interactive Math Toolkit), click on the curled page corner at the bottom right corner of your Interactive Math screen.

To Use the Calculator Tool:

To use the calculator tool in the Open Environment, click on the calculator icon in the main toolbar. Pull down the View menu at the top of the Calculator window, and choose either Standard or Scientific.

To Make a Spreadsheet:

1) Open the spreadsheet tool by either clicking on the spreadsheet (looks like a table) icon in the main toolbar, OR by pulling down the Toolbox menu at the top of the screen, selecting Data Analysis Tools, and then selecting Spreadsheet.

2) If the contents of a box or cell are too large for the size of the cell, then you can expand the width of the cell by dragging the right-hand border of the column's heading to the right. Column headings have letters in them whereas row headings have numbers.

To Make a Histogram:
(for *Open Environment Activity—What Is Average?*—Chapter 9, pages 85-86)

1) Open the histogram tool by pulling down the Toolbox menu, selecting Data Analysis Tools, and then selecting Histogram.

2) Then open a spreadsheet by using the directions above. Type in the data that you want to use in creating this histogram. Suggested data is the first column and last column of the frequency distribution table in part *a* of the activity.

3) Click on the Link tool in the primary toolbar, the first icon in the icon toolbar.

4) Click and hold the mouse down onto the Link icon in the title bar of the spreadsheet. It looks the same as the Link icon in the primary toolbar.

5) Drag the cursor to the Histogram window. A line connecting the two windows appears.

6) Release the mouse button. The data from the spreadsheet should now be displayed in the Histogram window.

To Make a Pie or Circle Graph

(for *Open Environment Activity—What Is Average?*—Chapter 9, on pages 85-86)

1) Pull down the Toolbox menu, drag down to Data Analysis Tools, and select Pie Graph.

2) Open a spreadsheet using the directions above. Type in the data that you want to use in creating this pie graph. In most cases, this will be percentages. Use the percentages of students with A's, B's, C's, D's, and F's as determined in parts *a* and *c* of the activity. Remember that the percentages should total to 100%.

3) Click on the Link tool in the primary toolbar. It is the first icon.

4) Click and hold down the mouse button on the Link icon in the title bar of the spreadsheet.

5) Drag the cursor to the Pie Graph window. A line connecting the two windows appears.

6) Release the mouse button. The data from the spreadsheet should now appear in the Pie Graph.

To Print Your Spreadsheet, Histogram, or Pie Graph:

Pull down the File menu at the top left corner of the screen, and select Print.

Chapter 1—The Whole Numbers
Key Concept Activity #1: More Bash for the Buck

1. Extension Exercise—Hidden Costs of Planning a Party

Your company is planning an employee holiday party. In addition to the direct costs, such as food and door prizes covered by the budget, the company president wants to know how much the party will cost the company in terms of planning time. You need to help the party committee calculate the hidden labor costs for planning the party. The four employees on the party committee are Karen, Tom, Juan, and Debbie.

a. After subtracting vacations and holidays for the year, Karen, Tom, Juan, and Debbie, each work 48 weeks a year, 40 hours a week. Determine how many hours each employee works in a year.

Hours worked per year per employee: ________

b. The annual salaries for Karen, Tom, Juan, and Debbie are shown in the table below. For each person, determine the hourly cost (in dollars) to the company, or how much one hour of work costs the company. Show your answers in the table.

Hint: hourly cost = yearly pay ÷ number of hours worked.

Employee	Yearly Pay ($)	Calculations	Hourly Cost to Company ($)
Karen	$28,800		
Tom	$30,720		
Juan	$32,640		
Debbie	$36,480		

c. Karen, Tom, Juan, and Debbie met five different times for two hours each to plan the party. Each meeting was held around lunchtime, so that one hour of the meeting was during the lunch hour (not on company time) and the other hour was on company time. How many hours of company time did each committee member spend planning the party? Show your work.

Hours of company time each committee member spent planning: ________________

d. In the chart below, copy the hourly cost for each employee (from part *b*). For each of the committee members, record the number of planning hours on company time (from part *c*), and determine how much the planning meetings cost the company in lost wages.

Employee	Hourly Cost ($) --copy from part *b*	Hours on Company Time --copy from part *c*	Calculations	Cost to Company ($)
Karen				
Tom				
Juan				
Debbie				

e. Explain why the direct costs of a party, such as food and drinks, decorations, and prizes do not necessarily cover all of the costs when a company plans a party for its employees. Explain what other hidden or direct costs the company might have if the party is held on company property or during regular work hours.

____

____

____

____

____

____

____

2. Open Environment Activity—Comparing Economic Data

The information in the chart below is taken from the U. S. Census Bureau. It compares economic information for three industries in the United States for the years 1992 and 1997.

Type of industry	Annual payroll (in dollars)		Paid Employees	
	1992	1997	1992	1997
Mineral	494,108,853	596,170,493	16,948,842	17,633,977
Retail	222,867,879	293,577,652	18,407,453	21,349,109
Service	639,369,976	921,942,255	27,399,296	34,223,932
Total				

a. For 1992 and 1997, find the total annual payroll (in dollars) and the number of paid employees in the mineral, retail, and service industries. Record results in the table above. Use the vertical format of the numbers to help you calculate the totals for each column.

b. Now determine the increase in annual payroll from 1992 to 1997 and the increase in the number of paid employees from 1992 to 1997. You can complete this part by hand or by using the scientific calculator in the Open Environment. Write your results in the table below. *Hint:* The increases are actually differences.

To use the calculator, enter the Open Environment (also called Interactive Math Toolkit) by clicking on the curled page corner at the bottom right corner of your Interactive Math screen. Once you are in the Open Environment, click on the calculator icon. Pull down the View menu at the top of the Calculator menu, and click "scientific." You can use this scientific calculator to find the increases.

Type of Industry	Increase in Annual Payroll (in thousands of dollars)	Increase in Number of Paid Employees
Mineral		
Retail		
Service		

c. Let's learn how to build a spreadsheet by creating the table in part *b*. We will use the spreadsheet tool in the Open Environment. Follow the steps below. You will also find these directions on pages ix-x of this workbook.

1.) Open the spreadsheet tool by either clicking on the table/spreadsheet icon in the toolbar, OR by pulling down the Toolbox menu at the top of the screen, selecting Data Analysis Tools, and then Spreadsheet.

2.) Expand the size of the spreadsheet by pulling on its outer border to the right or pulling on the bottom right corner of the spreadsheet.

3.) Expand the width of a column by pulling on the border of its heading (represented by the letters A, B, C, etc.) and pulling its outer margin to the right.

4.) Type the heading "Type of Industry" in row #1, column A. Expand the size of the box or cell as needed by following the directions in step #3.

5.) Do the same for the headings "Increase in Annual Payroll ($)" and "Increase in Number of Paid Employees." You will use row #1 and columns B and C .

6.) Now in column A, you will add the row headings: "Mineral," "Retail," and "Service" in rows 2, 3, and 4 respectively.

7.) Keep this spreadsheet open and fill in the table using your results in part *d* below.

8.) Print your spreadsheet by pulling down the File menu at the top left corner of the screen, and selecting Print.

d. For each type of industry in your table from part *b*, round the increase in annual payroll to the nearest million dollars. Then round your data for the increase in number of paid employees to the nearest hundred thousand. Complete your spreadsheet from part *c* by filling in these results, and print the final spreadsheet.

e. Describe the process that you would use to find the average salary for each industry in 1992. Which data would you use? Determine whether or not your results would be whole numbers.

____

____

____

____

3. Conceptual Exercise—Safe Dose?

Claudia, a 7-year old girl, weighs 19 kilograms. The manufacturer's recommended safe range for doses of heparin is 10 to 25 units per kilogram per hour. Through a series of steps, you will determine the safe range of doses for Claudia and determine her dose of heparin per hour.

a. **10 units of heparin per kilogram per hour is the low end of the recommended safe dose. What is the lowest safe dose of heparin that should be given to Claudia in one hour?**
Hint: **The lowest safe dose (number of units) of heparin for Claudia in one hour = the lowest safe number of units of heparin recommended per kilogram per hour × Claudia's weight in kilograms.**

Lowest safe dose of heparin for Claudia in one hour: ______________________ units

b. **25 units of heparin per kilogram per hour is the high end of the recommended safe dose. Calculate the highest safe dose of heparin per hour for Claudia.**

Highest safe dose of heparin for Claudia in one hour: ______________________ units

c. **To summarize the results in parts *a* and *b*, record the recommended safe range (low to high) of units of heparin per hour for Claudia.**

From ______________________ to ______________________ units of heparin per hour

d. Claudia must receive her dose of heparin as part of an IV or intravenous fluid. The total amount of the IV solution (containing her dose of heparin) is 250 milliliters, and it flows at a rate of 50 milliliters per hour. How many hours will it take to administer 250 milliliters of the intravenous (IV) solution containing heparin? Show your work.

________________ hours

e. Suppose that the total IV solution (250 milliliters) contains 2000 units of heparin. How many units of heparin are administered to Claudia per hour? Show your work.

Hint: The number of units of heparin administered to Claudia per hour = total number of units of heparin in the entire 250-milliliter solution ÷ the number of hours it would take to administer the entire solution.

________________ Units of heparin administered to Claudia per hour

f. Compare your answer in part *e* to the safe range of dosages for Claudia as determined in part *c*. Is this dose within the recommended safe range? Explain why or why not.

____

____

____

____

____

4. Group Activity—Where Did All The Money Go?

Mary Thurmond willed her entire estate to be divided equally among her 22 nieces and nephews. When she died, her estate was valued at over two million dollars. First, in a very simplistic world where there are no fees taken out of the estate, you will determine the amount of inheritance for each of the nieces and nephews. Next, you will determine how to distribute the funds after taxes and probate (the act of verifying the will's validity) fees are subtracted from the estate.

Work with your group members to answer the following questions.

a. In addition to her residence, Mary Thurmond owned a farm that was leased to a corporate farmer, and a beach house. Her real estate properties were appraised as shown in the chart below. Calculate and record the total appraised value of her real estate properties in the table.

Property	**Appraised Value**
Residence	$435,000
Beach House	$350,000
Farm	$730,000
Total	

b. An estate auctioneer valued her personal belongings at $115,000. Investments in money market accounts and stock market accounts are estimated at $573,000. Find the total estimated value of her personal (not real estate) property and investments.

c. What is the total estimated value of all of Mary Thurmond's properties and investments? Round this amount to the nearest hundred thousand dollars.

d. Work with your group members to calculate each nephew's and niece's inheritance based on your final estimate in part *c*.

e. An auction was held to sell Mary Thurmond's personal property. The real estate properties and stocks were sold, and the money market accounts were closed out. All of this money was held in an escrow account. (Note: An escrow account is money or property placed into the hands of a third party, and is delivered to the grantee only after fulfilling specific conditions.)

The money collected after all real estate, broker, and auctioneer fees were paid is shown in the table below. Find the difference between the actual dollar amount collected after fees and the appraised value. Then determine whether the actual dollar amount after fees reflects a gain or loss in value from the appraised estimate.

Properties	**Actual Amount Collected After Fees**	**Difference from Appraised Value**	**Gain or Loss in Value?**
Residence	$394,190		
Beach house	$360,100		
Farm	$522,000		
Personal property	$53,250		
Stocks and Money Market	$427,930		
Total			

What is the total amount of money that was deposited into the escrow account?

____

f. Calculate the share of the "total in escrow" for each nephew and niece. Find the difference between this share and the estimated share based on appraised values (from part *d*).

g. Probate court fees came to $55,000, and taxes totaled $575,230. What total sum was left to divide among the heirs after these costs are paid from the escrow account?

h. After taxes and probate fees are paid, how much money will finally be given to each nephew and niece?

i. Calculate the difference between what each niece and nephew might have expected given the appraised values and their actual share.

Chapter 2—Multiplying and Dividing Fractions
Key Concept Activity #2: Cookie Cutting

1. Extension Exercise—Space Costs

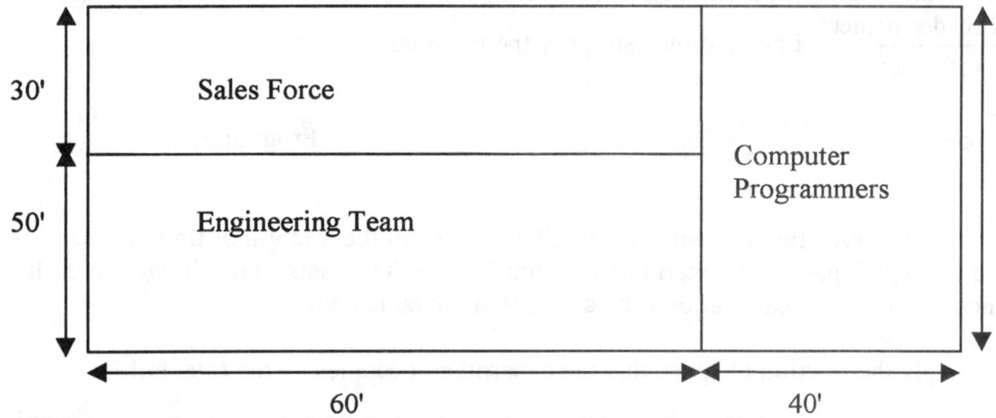

For internal bookkeeping purposes, many companies charge costs such as salaries and supplies to each department and also split operating expenses among the departments based on the amount of space that each department occupies.

In this exercise, you will find the floor-space area of the entire building as well as the floor-space area of each department. Then you will determine what fraction of the building each department occupies and use that information to determine how to divide expenses. Length and width are measured in feet. Area is measured in square feet.

a. Find the length and width of the entire building. Show your calculations and include correct units with your answers.

Length = ____________________ Width = ________________________

b. What is the total area of the entire building? Show your calculations and include correct units with your answers. *Hint:* area = length × width

Area = ________________________

c. Calculate the area of each department in the building.

Sales: _____________ Engineering: ____________ Programming: _______________

d. Using the total area of the building and the area of each department from parts *b* and *c*, determine what fraction of the total area that each department occupies. Write the fractions as $\frac{\text{area of department}}{\text{total area}}$ first, and then simplify the fractions.

Sales force: ____________ Engineering: ______________ Programming: _________

e. The company received the following utility bills. Based on the fraction of floor-space occupied by each department, determine how much of each bill should be charged to each department. Show your calculations and answers in the table below.

Hint: Multiply the fraction of space that the department occupies by the total bill.

Department	**Share of Electric Bill** Total bill = $2760	**Share of Gas Bill** Total bill = $760	**Share of Water Bill** Total bill = $280
Sales Force			
Engineering			
Programming			

f. The company made improvements to the property totaling $183,600. This dollar amount loses value or depreciates over a 10-year period, meaning that $\frac{1}{10}$ of the amount depreciates each year. By how many dollars does the amount depreciate each year?

g. If the improvement costs were divided among the departments according to the amount of floor space they occupy, how much would be charged to each department every year for ten years? Show your calculations.

Sales force: _______ Engineering: __________ Programming: ____________

2. Open Environment Activity—Bills and Coins

In the United States, the value of a coin or bill is based on a fraction or multiple of the basic unit, the dollar. A half-dollar and a quarter are so named because they are ½ of a dollar and ¼ of a dollar, respectively.

a. In the table below, fill in how many coins it takes to equal one dollar and what fraction of a dollar (or 100 cents) each coin is. The first row has been completed for you.

Coin	Number of Coins Needed to Make $1	One Coin is What Fraction of a Dollar?	Simplify the fraction.
Penny	100	$\frac{1}{100}$	$\frac{1}{100}$
Nickel			
Dime			
Quarter			
Half-dollar			

b. Three quarters = _________ pennies

c. Using the information from part *a*, write the fraction of a dollar that represents 75 pennies. Now write the fraction of a dollar that represents three quarters. Show that the two fractions are equal using cross products.

d. One half-dollar has the same value as how many pennies? nickels? dimes? quarters?

1 half-dollar = ______ pennies = _____ nickels = ______ dimes = _______ quarters

Now represent this series of equal values as equivalent fractions. The first fraction is written for you. Prove that all are equivalent fractions by finding the cross products of the first two fractions, the cross products of the second and third fractions, and so on.

$\frac{1}{2} =$

Vanessa works on the floor of a casino selling tokens and chips to customers. The tokens come in rolls, and the rolls are put in *squares* of 25 rolls. A *bucket* is equal to two squares.

e. Using the directions for making a spreadsheet on pages ix-x of this workbook, build a spreadsheet like the table below to determine the value of one roll of each type of token, the value of one square of tokens, and the value of one bucket of tokens. The first row has been completed for you. Write your answers in your spreadsheet and in the table below. Turn in a printed copy of your spreadsheet when you submit this activity to your instructor.

Type of Token	**Value of Token ($)**	**No. of Tokens in a Roll**	**Value of one Roll ($)**	**Value of one Square (25 rolls) ($)**	**Value of one Bucket (2 squares) ($)**
H	$\frac{1}{2}$	40	$20	$500	$1000
Q	$\frac{1}{4}$	40			
T	$\frac{1}{10}$	100			
N	$\frac{1}{20}$	100			

f. Vanessa invested her hard-earned money and bought 480 shares of stock at $7\frac{3}{8}$ dollars per share. How much did the 480 shares cost? Show your work.

g. In the saying, "Two bits, four bits, six bits, a dollar," follow the pattern to determine how many bits are in one dollar and the value of 2 bits, 4 bits, 6 bits, and 8 bits as a fraction of a dollar. How many quarters equals 2, 4, 6, and 8 bits?

Number of bits in a dollar: ____________

No. of Bits	**Fraction of a Dollar**	×	**Number of Quarters in a Dollar**	**Number of Quarters Equal to 2, 4, 6, or 8 Bits**
2		×	4	
4		×	4	
6		×	4	
8		×	4	

3. Conceptual Exercise—Correct Dose?

a. How many $\frac{1}{16}$-milliliter capsules make up one milliliter of morphine? Draw capsules to illustrate your answer.

b. Morphine at the hospital comes in $\frac{1}{16}$-milliliter capsules. A single dose of morphine equals $\frac{1}{32}$ milliliters. In terms of capsules, how much morphine is in a single dose?

One dose = _______ capsule(s)

c. Dr. Al-Hamar says the patient can have up to $\frac{1}{2}$ milliliters of morphine a day. If the patient takes the maximum daily amount, how many $\frac{1}{32}$-milliliter doses should the patient receive in 24 hours? Show how you arrived at your answer, and verify that the total medication in 24 hours is $\frac{1}{2}$ milliliters.

Number of $\frac{1}{32}$-milliliter doses in 24 hours: ________

d. If the patient receives $\frac{1}{2}$ milliliters of morphine a day and the $\frac{1}{32}$-milliliter doses are evenly spaced over 24 hours, then how many hours apart are the doses? Write your answer as a mixed fraction.
Hint: 24 hours ÷ number of doses a day = number of hours between doses

Number of hours between doses: ______________

e. Suppose that Dr. Al-Hamar prescribes $\frac{1}{32}$ milliliter doses of morphine as the patient needs it, but is careful not to administer doses any closer than 2 hours apart. What would be the maximum number of doses possible in a day? What would be the maximum amount of morphine administered in one day?

Maximum number of doses in 1 day: _______ doses.

Maximum amount of morphine: ______ milliters.

f. Another drug comes in $\frac{1}{16}$-milliliter capsules. Dr. Al-Hamar prescribes a single dose of $\frac{1}{8}$- milliliter. Draw capsules to illustrate how many capsules make up one milliliter of morphine. Then, show how much $\frac{1}{8}$ milliliter of morphine would be in terms of capsules.

Number of capsules in 1 milliliter of morphine: __________
There are ______ capsule(s) in one dose.

g. If the doctor's prescription says to administer a single dose of $\frac{1}{8}$ milliliter every four hours, then how many doses are to be given in one day (24 hours)? How many capsules are given in one day? Show your calculations.

Number of doses: _________ per day
Number of capsules ________ per day

h. If a doctor's prescription says to administer one dose of $\frac{1}{8}$ milliliter of a drug every four hours for three days, how many doses and how many capsules will be taken in the 3-day period?

Number of doses in 3 days: _______ Number of capsules in 3 days: ____________

4. Group Activity—Cubes and Fractions

Materials Needed:

- Paper to make 2 squares--one with a side length of ½ inch to represent a ½-inch cube, and the other with a side length of ¾ inch to represent a ¾-inch cube
- ruler

A cube is a three-dimensional object shaped like a sugar cube where length, width, and height are the same measurement. A cube that is ½ in. x ½ in. x ½ in. is called a half-inch cube (½-inch cube). A cube that is ¾ in. x ¾ in. x ¾ in. is called a three-quarter-inch cube (¾-inch cube). You and your group members will use paper squares to represent one side of a ½-inch cube and one side of a ¾-inch cube.

Work with your group members to answer the following questions.

a. If you line up 13 half-inch-cubes side by side, as illustrated below, then what is the total length of the cubes together? Use a ruler and your ½-inch paper cube (square) to measure the length of 13 half-inches.

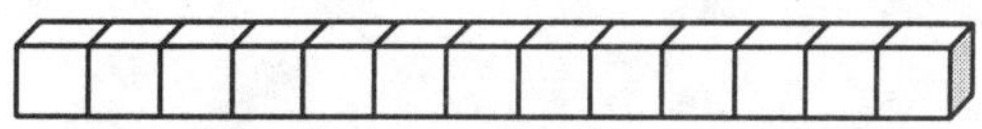

Length of 13 half-inch cubes ________ inches

b. How would you use multiplication to find the total length of 13 half-inch cubes? Express the length both as an improper fraction and as a mixed number. Show your work.

________________________ = ______________ inches
(improper fraction) (mixed number)

c. If you line up 16 three-quarter-inch cubes side by side, then what is the total length of the cubes together? Use a ruler and your ¾-inch paper cube (square) to measure the length of 16 three-quarter-inches.

d. How would you use multiplication to find the total length of 16 three-quarter-inch cubes? Show your answer as an improper fraction, and then write your answer in simplest form.

Improper fraction: ______________ inches **Simplest form:** ____________ inches

How would you build a box with ½-inch cubes? The bottom layer of the box and the entire box are pictured below.

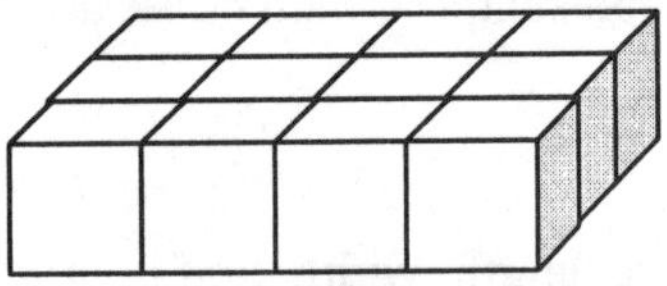

Bottom layer of the box has 3 rows of half-inch cubes and 4 columns or 4 cubes in each row

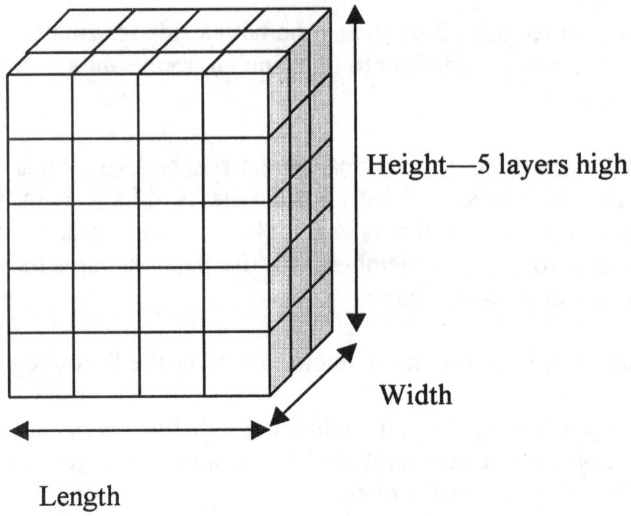

One half-inch cube

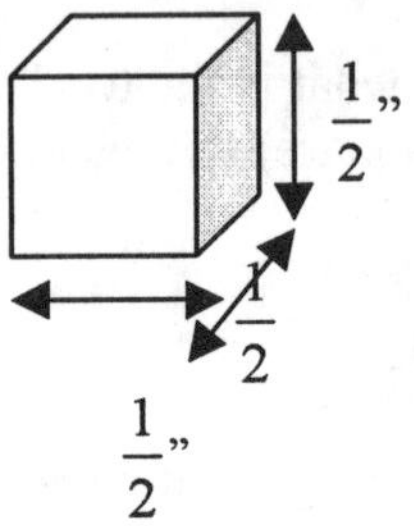

e. If you stack the box five layers high, find the length, width, and height of the resulting box. Use the box illustration above to help you find the answer. Express each measurement in terms of number of cubes and in terms of number of inches. If a measurement does not result in a whole number of inches, then express your answer as an improper fraction and as a mixed number.

Length: ______________ cubes = ____________________________ inches

Width: ______________ cubes = ____________________________ inches

Height: ______________ cubes = ____________________________ inches

f. The volume of a rectangular box is found by multiplying length × width × height of the box. Calculate the volume of the 5-layered box above using the inch measurements that you found in part *e*. Write the volume in simplest form. If the result is not a whole number, then express your answer as both an improper fraction and a mixed number.

Volume = ______ x _________ x _________ = ________________ cubic inches
(length) (width) (height)

Chapter 3—Adding and Subtracting Fractions
Key Concept Activity #3: Playing the Market

1. Extension Exercise—Monitoring Your Stocks

Suppose you decide to go on-line to check the stock that you own in different companies. First you find the following data on Dover Computer Corporation.

Ticker Symbol	Today's High	Today's Low	Last Price	Prior Close	Change	Trade Time
Dover	$42\frac{15}{16}$	$39\frac{1}{2}$	$41\frac{3}{8}$	$43\frac{9}{16}$	$\downarrow 2\frac{3}{16}$	12:29 p.m.

a. As of 12:29 p.m. this trading day, what is the difference between today's high price per share and today's low price per share? Show all your work.

b. The last price represents the price of a share as of 12:29 p.m., and the prior close was the price per share at the close of trading on the previous day. Show how subtracting the last price from the prior close price yields the difference of $2\frac{3}{16}$. Then explain what the arrow indicates in the "Change" column.

c. If you own 1000 shares of Dover, then how much value did your Dover investment lose during the time from the prior closing until 12:29 p.m. today? Do you think this is a cause for concern? Explain.

---

d. Now observe data from another company that you own stock in. What must the "last price" be in order to obtain the "change" of $\uparrow 2\frac{19}{32}$?

Ticker Symbol	Last Price	Prior Close	Change	Trade Time
Medicon	?	$5\frac{1}{4}$	$\uparrow 2\frac{19}{32}$	12:29 p.m.

e. If you own 9600 shares of Medicon, then how much value did your Medicon investment gain during the time from the prior closing until 12:29 p.m. today? Do you think this is a cause for celebration?

---

2. Open Environment Activity—Cooking Dinner

Tonight it is your turn to cook dinner, and you need to throw something together quickly. Steak? Fish? Chicken? You're not sure, but you know that everyone likes rice. So you grab a pan and look on the back of a Ready-in-a-Minute brand of rice. There you see a measurement table similar to the one below.

Servings	1	2	4	5	6	9
Rice (cups)		$\frac{1}{2}$	1		$1\frac{1}{2}$	
Water (cups)		$1\frac{1}{4}$	$1\frac{7}{8}$		$3\frac{1}{8}$	
Salt (teaspoons)		$\frac{1}{4}$	$\frac{1}{2}$		$\frac{3}{4}$	
Butter or Margarine (teaspoons)		1	2		3	

a. What value could you multiply the 2-serving column by to convert to 1-serving?

b. Fill in the table above for 1-serving. Be sure to simplify your answers. Then make this table into a spreadsheet in the Open Environment. Follow the directions for making a spreadsheet on pages ix-x of this workbook.

c. Using the 1-serving and 4-serving columns, how would you figure out the 5-serving column using addition?

____

____

d. Fill in the 5-serving column on your spreadsheet (or table) by using the procedure that you just described. Express your results as a mixed number when appropriate, and make sure that any fractions are written in simplest form.

e. How many servings can you make if you have two cups of rice left and want to use all of it?

Calculation:

Answer:

f. How could you use addition to find the proper amounts for 9 people? Which columns would you use?

____

____

g. In your spreadsheet (or table), fill in the column for 9 servings using addition. Be sure to simplify any fractions and write your answer as a mixed number when appropriate.

h. How could you find the proper amounts for 9 servings using fewer steps and one operation?

____

____

i. What process is easiest to use if you want to find a large number of servings?

____

j. Use the process that you described in part *i,* find the proper amounts of each ingredient for 72 people. Write your answers in the table below.

Servings	72
Rice (cups)	
Water (cups)	
Salt (teaspoons)	
Butter or Margarine (teaspoons)	

3. Conceptual Exercise—Team Effort

Doug is tired of spending three hours every Saturday morning mowing the lawn while his son, Michael, watches cartoons on TV. The last time that Michael mowed the lawn it took him three hours, and the job was only half done. Still, Doug figures that if he borrows his neighbor's lawn mower and they work together, this will give his son a sense of accomplishment and a positive work ethic.

a. Since Doug is able to mow the whole lawn in 3 hours, what fraction of the lawn can he mow in 1 hour?

b. If Michael can mow half the lawn in 3 hours, then how long will it take him to mow the whole lawn alone?

c. Using your result from part *b,* what fraction of the lawn can Michael mow in 1 hour?

d. Working together, what fraction of the lawn can Doug and Michael mow in 1 hour? Show all work including the process of finding a least common denominator. Write your answer in simplest form.

e. How long does it take to mow the whole lawn with father and son working together?

f. Why is it helpful to find the fraction of the lawn each person could mow in 1 hour?

____

____

____

____

g. The concept of finding a common denominator can be visualized with a geometric model. Let the whole lawn be represented by the rectangle below. Split the rectangle into the appropriate number of parts to represent how long it takes Doug to mow the entire lawn. Then shade in the portion of the rectangle to show the part of the job that Doug completes in 1 hour working alone.

h. Now, let the same lawn be represented by another large rectangle given below. Split the rectangle into the appropriate number of parts to represent how long it takes Michael to mow the entire lawn. Shade in the portion of the rectangle to show the part of the job that Michael completes in 1 hour working alone.

i. If Doug and Michael work together, then you can add the fractional amount that each mows in 1 hour separately. Show this geometrically by placing your two models from parts *g* and *h* below and find their sum.

Doug Michael

+ =

j. If you draw a horizontal line through the middle of Doug's rectangle, this demonstrates the equivalence of what two fractions?

k. What do your answers in part *i* and part *d* have in common?

____

____

4. Group Activity —Pizza! Pizza!

On pizza night, Jim and his seven friends are gathered in their dorm room for pizza and TV. The group orders four whole pizzas: one mushroom, two pepperoni, and one anchovy. When the pizza arrives, it doesn't take long for the whole mushroom pizza and one of the pepperoni pizzas to disappear. After all have eaten their fill, there are 2 slices of pepperoni and 3 slices of anchovy left.

Work with your group members to answer the following questions.

a. Draw two circular pizzas. Label one "anchovy" and the other "pepperoni." Now, divide each pizza into 8 equal parts, and shade-in the parts that represent the remaining pizza.

b. Use a fraction to represent the amount of pepperoni pizza remaining out of that whole pizza. Then do the same with the remaining amount of anchovy pizza. Do not simplify your fractions yet.

Pepperoni fraction:

Anchovy fraction:

c. Combine the pepperoni fraction with the anchovy fraction to obtain the total amount of a whole pizza leftover.

d. How much pizza have Jim and his friends consumed so far? State your answer in terms of an improper fraction and a mixed number. *Hint:* The numerator of the improper fraction should give you the number of slices consumed, and the mixed number will give you the amount consumed in terms of whole pizzas.

Improper fraction: Mixed number:

e. Three members of the group are interested in the leftover pizza. How many slices do you have to split three ways? What is the fractional part of one whole pizza that is available for one group member? *Hint:* Remember that there are 8 slices in one whole pizza.

____

____

f. Is your answer to the last question in part *e* likely to be the way that the three people divide the pizza? Why or why not?

____

____

g. Just when everything is settled on the leftovers, you turn to leave and knock one of the pizza slices on the floor. Assuming no one wants that piece, what fractional part of one whole pizza is now available to divide three ways? Write your answer in simplest form.

h. Suppose that a student from the next dorm room brings in another $1\frac{1}{4}$ pizza that was leftover from his dorm party. Using your answer from part *g,* how much pizza is now available to split three ways evenly? Write this answer as a mixed number.

Chapter 4—Decimals
Key Concept Activity #4: Going the Distance

1. Extension Exercise—A Year at MIU

Suppose you have just started as an exchange student at Maritime International University (MIU). The apartment where you will live is 10 kilometers from campus.

a. If you take your car from your apartment to the MIU campus, how does travel time depend on average speed? State hours in simplest fractional form. Then give minutes in either whole numbers or in decimal form rounded to the tenth's place.

$$\text{Time} = \frac{\text{Distance}}{\text{Rate}}$$

Distance (km)	Rate (km/hr)	Time (hours)	Time (minutes)
10	10	1	60
10	20		
10	36		
10	40		
10	45		
10	55		
10	60		
10	70		
10	80		

b. Examine the second and third columns. Every time you double the average speed, what happens to the trip time? Answer in complete sentences using examples to support your conclusion.

___

___

___

___

c. Suppose your average walking pace is 2.7 kilometers per hour. How long will it take you to go from your apartment to the MIU campus? Show all work and present your answer in decimal form. Then convert this time into minutes. Round all times to the nearest tenth's place.

Your car has a mileage of 25.6 kilometers per gallon when traveling at an average speed of 40 kilometers per hour. Your car's gas tank holds 16.5 gallons.

d. If you are driving at a constant rate of 40 kilometers per hour, what is the total distance (in kilometers) that your car can travel on one tank of gas if all the gas is used?

e. Suppose the price of gasoline is equivalent to $1.99 per gallon. What is the cost of filling an empty tank? Round your answer to the nearest cent.

f. What is the cost of gas per kilometer if your speed is 40 kilometers per hour? Round your answer to the nearest cent.

g. If your first class is at 8 a.m. and you drive at an average speed of 40 kilometers per hour, then at what time should you leave for school each morning? Explain why there might be more than one good answer to this question.

h. If you use your car to travel from home to work each day with no side trips, then how often should you fill your car's gas tank? Round your results to the nearest whole number. *Hint:* You will use your answer from part *d.*

i. If the cost of gasoline stays fixed at $1.99 per gallon for the entire 15-week semester, then about how much money will fuel cost over the entire semester? Assume that you do not drive on weekends and that classes occur during the 5-day week. Round results to the nearest cent.

2. Open Environment Activity—Exchange Rates

In May of 1999, the rate of exchange between the United States dollar and selected foreign currency was as listed in the table below. Exchange rates can tell you what U.S. dollars are equivalent to in foreign currency or what a foreign currency is equivalent to in U.S. dollars. For example, 1 U.S. dollar equals a little over 6 French francs, about 121 Japanese yen, or 1800 Italian lira. The values in the table list the exchange rates as units of foreign currency per $1.

The franc/dollar exchange rate can be expressed as $\frac{6.10 \text{ francs}}{1 \text{ dollar}}$, meaning that there are 6.10 francs to every dollar. This rate can also be written as $\frac{1 \text{ dollar}}{6.10 \text{ francs}}$, meaning that there is 1 dollar to every 6.10 francs.

a. Suppose you buy a basic TV for $500 in the U.S. The third column expresses how many foreign units there are per $1. Specify in the 4th column the price of the same TV using the currency of each country. Complete the table below, OR make a spreadsheet of the table in the Open Environment using the directions on pages ix-x of this workbook.

Country	Currency	Rate Units per $1	Foreign Currency Price of $500 TV	Rate in U.S. $
France	Franc	6.10		
Britain	Pound	0.61		
Germany	Mark	1.82		
Italy	Lira	1800		
Japan	Yen	120.50		
Canada	Dollar	1.45		
Mexico	Peso	9.27		

(Exchange Rates recorded on May 5, 1999)

b. If a Canadian want to buy a car in the U.S. that sells for $20,000, how many Canadian dollars must be exchanged to make the purchase in U.S. dollars? Show all work.

c. If a Mexican crosses the border to earn $8.50 an hour, how many pesos can be exchanged upon returning to Mexico after working 40 hours in the U.S.? Assume there are no deductions from the paycheck and that all dollars earned will be exchanged at the above rate. Show all work.

d. Use the fifth column in the table or your spreadsheet from part *a* to approximate each of the exchange rates as U.S. dollars per unit of foreign currency. For example, 1 U.S. dollar per 6.10 francs can be expressed as the ratio, $\frac{1 \text{ dollar}}{6.10 \text{ francs}} \approx .1639 \text{ dollars / franc}$, meaning there are 0.1639 dollars in 1 franc or 1 franc is equal to 0.1639 dollars (about 16% of a dollar). Use the calculator in the Open Environment as needed. Directions for using the calculator appear on pages ix-x of this workbook. Round results to the thousandth's place. For the lira, round results to the ten-thousandth's place.

e. If you want to take a 27,672.30 franc vacation on the French Riviera, how many U. S. dollars must you exchange for the trip?

f. Suppose you have the opportunity to take a job in Japan that has a salary of 7,225,800 yen. How many U.S. dollars would you have to be paid to make the same amount in this country?

The table in part *a* is set up for the conversion between U.S. dollars and foreign currency. However, you can also calculate the exchange rate of each foreign currency in terms of any other foreign currency. For example, we know that 1 U.S. dollar equals 6.10 francs (f), and 1 U.S. dollar equals 0.61 British pound (£). This implies that 6.10 francs equals 0.61 pounds.

This is an example of cross-exchange rates. The franc/pound exchange rate can be calculated by dividing both sides of the equality by 0.61,

$$\frac{6.10f}{0.61} = \frac{0.61£}{0.61}$$

$10f = 1$ £ or 10 francs equals 1 pound.

g. Use the ideas developed above to complete the following table of cross-exchange rates. Round to the thousandth's place. In the sixth column, round units per 1 lira to the ten-thousandths place.

Currency	units per 1 dollar	units per 1 franc	units per 1 pound	units per 1 mark	units per 1 lira	units per 1 yen	units per 1 peso
dollar	1	0.164	1.638	0.549	0.0006	0.008	0.108
franc	6.10	1					
pound	0.61		1				
mark	1.82			1			
lira	1800				1		
yen	120.50					1	
peso	9.27						1

3. Conceptual Exercise—Traveling for Big Brother

You have been asked to be a consultant for the Big Brother Oil Corporation's mining and coal division. Your task is to examine the most economical way to transport coal to an electric company.

a. Suppose preliminary research shows three possible methods of transportation.

- Trucks traveling 150 miles to the destination
- Freight trains covering 800 miles of railroad to reach the destination
- A barge going for 200 miles on water, then a freight train covering 700 miles of railroad

Do you have enough information to make a recommendation to Big Brother? If yes, go to the next page and write a brief report on the best course of action. Otherwise, state what additional things must be known to make an informed decision.

____

____

____

b. More research reveals the following data on the cost of transporting coal.

- The truck costs $0.17 per ton per mile.
- The freight train costs $0.03 per ton per mile.
- The barge costs $0.02 per ton per mile.

Based on the information gathered so far, which method would you recommend to minimize the cost of moving the coal? Show all work below, then in part *d* write a report to Big Brother Oil Corporation justifying your conclusions.

c. Use the data below to answer this question.

- The train travels at 70 miles per hour.
- The barge travels at 25 miles per hour.
- The truck travels at 55 miles per hour.
- The train conductor makes 3 stops for a ½ hour break at each stop.
- The trucker makes 3 stops of 15 minutes each.
- The barge conductor travels straight through.
- $\text{Time} = \dfrac{\text{distance}}{\text{speed}}$

Assuming that all of the data above is true, which of the routes listed in part *b* will take the least amount of time? Show all of your work, and round your time units to the nearest tenth of an hour.

d. Write a brief report to Big Brother Oil Corporation explaining the most economical way to transport coal to the electric company. Is this the best choice for time reasons? Use your answers from parts *b* and *c* to explain your choice. Give both the pros and cons of your decision.

___

___

___

___

___

___

___

___

___

___

___

4. Group Activity —The Donut Dilemma

The alarm jolts you awake on a Monday morning. It is your turn to bring donut holes to your 8:00 a.m. math class. The local donut shop sells donut holes in boxes of 20, 45, and 60. You have $10 of spending money that must last until Friday, and you are wondering what to purchase at the donut shop. Work with your group members to discuss and answer each of the questions.

a. List what questions you need to consider before you buy any donut holes.

___

___

___

___

b. Suppose the following data is known:

- One box of 20 donut holes cost $1.99 including tax.
- One box of 45 donut holes cost $3.29 including tax.
- One box of 60 donut holes cost $3.69 including tax.
- There are 33 students registered for your math class.
- On average, 5 students are absent from any class meeting.
- On average, each student eats 3 donut holes.

How many boxes of each size should you buy to make sure that everyone has enough to eat, while at the same time minimizing the cost? Show all work and justify your answer in words using complete sentences.

___

___

___

___

c. Explain why part *b* has more than one right answer.

d. What is the maximum number of donut holes that can be bought for $7.00? Show all work and justify your answer in words using complete sentences.

e. How many correct answers are there to part *d*? In what way is this question different from part *b*? Explain.

Chapter 5—Ratio and Proportion
Key Concept Activity #5: By Popular Vote

1. Extension Exercise—1996 Election Polls

At several times prior to the 1996 U.S. presidential election, the Gallup organization ran a series of polls where people were asked the following question:

> **POLL Question:**
>
> If the 1996 presidential election were held today with Bill Clinton as the Democratic candidate and Bob Dole as the Republican candidate, who would you be more likely to vote for—
>
> Clinton, the Democrat
> OR
> Dole, the Republican?
>
> If you are UNDECIDED, answer this question: As of today, do you lean more toward Clinton, the Democrat OR Dole, the Republican?

The following chart records the results of those polls at various dates. 100 people were polled in each case.

Dates of Poll	Registered Voters for Clinton	Registered Voters for Dole	Undecided
June 5 - 6, 1995	46	51	3
Jan 12 - 15, 1995	48	49	3
May 9 - 12, 1996	58	38	4
Aug 14 - 15, 1996	52	42	6
Aug 16 - 18, 1996	51	46	3
Nov 4 - 5, 1996	57	37	6

a. Use the data above to find the following:

Date of poll when Dole had his greatest support: ______________________________

According to that poll, who would have won the election? ______________________

b. 96, 389, 818 people voted in the actual 1996 presidential election. At the time when Dole had his greatest support, how many people out of the 96,389,818 voters would have voted for each candidate? Set up proportions and show your calculations below. *Hint:* Refer to the table on the previous page and your answer to part *a*. You should use the letter *n* to represent the unknown number and then cross multiply. Round your results to the nearest whole number.

Dole: proportion: _____ = ______ ____________ # of voters

Clinton: proportion: _____ = ______ ____________ # of voters

c. A similar poll was conducted on the day before the election. The responses out of 100 people are shown in the table below. Find the actual number of people who would have voted for each ticket if the election were held on that day. In the third column, show how you set up your proportions and the steps you used to solve the equation. Round results to the nearest whole number.

Party Ticket	**Poll Responses (out of 100)**	**Proportion**	**Number of Voters (out of 96, 389, 818)**
Clinton/Gore	52		
Dole/Kemp	41		
Perot/Choate	7		

d. Out of the 96,389,818 voters, the actual 1996 election results are 47,401,898 for Clinton, 39,198,482 for Dole, and 7,137,235 for Perot. Compare your answers in part *c* to the actual number of votes for each candidate. How well did the polls predict the outcome of the election for each candidate?

____

____

____

2. Open Environment Activity—A Trip to the Ozarks

Marc is planning a one-week vacation to the Ozarks to canoe, hike, and study geological sites of the area. He is considering renting a Honda Civic, a Ford Taurus, or a Jeep Cherokee but wants to compare rental and gasoline costs before making his decision.

The chart below shows the expected gas mileage, rental, and insurance costs for each of the three vehicles.

	Honda Civic	**Ford Taurus**	**Jeep Cherokee**
Rental Costs ($ per day)	$38	$40	$58
Insurance ($ per day)	$16.99	$16.99	$16.99
Expected Mileage on curvy roads/small towns (mpg)	32	19	20
Expected Mileage on Highway (mpg)	37	29	21

Marc estimates that he will drive a total of about 700 miles during his trip. 230 of those miles will be on the interstate, and the rest will be on curvy, windy roads or in small towns.

a. Using the data above, find the total number of gallons of gasoline needed for highway driving as well as for curvy-road driving in each of the vehicles.

Hint: Write each mileage as a fraction, $\frac{\text{number of miles}}{\text{1 gallon}}$, and set that fraction equal to the fraction, $\frac{\text{\# miles traveled}}{\text{\# of gallons needed}}$, to make a proportion. Fill in the known values and use cross multiplication to find the unknown value. Round results to the nearest tenth.

Car	**Curvy Road Proportion**	**Number of gallons for curvy road driving**	**Highway Proportion**	**Number of gallons for highway driving**
Honda Civic				
Ford Taurus				
Jeep Cherokee				

b. At a gasoline price of $1.12 per gallon, calculate the cost of gasoline for each vehicle. Then calculate the rental and insurance costs for each vehicle for seven days. Record your answers in the chart below. Round all results to the nearest cent.

Hint: Gas cost = cost of gas per gallon × number of gallons

Honda Civic		**Ford Taurus**		**Jeep Cherokee**	
Gas Cost (Curvy road)	Gas Cost (Highway)	Gas Cost (Curvy road)	Gas Cost (Highway)	Gas Cost (Curvy road)	Gas Cost (Highway)
Total Cost of Gas:		Total Cost of Gas:		Total Cost of Gas:	
Rental and Insurance Costs per week:		Rental and Insurance Costs per week:		Rental and Insurance Costs per week:	

c. Which car is the least expensive to rent and drive? Which car costs the most? Other than cost, what do you think Marc should consider in making his decision about which car to rent?

____

____

____

____

____

____

d. Use the directions on pages ix-x of this workbook to display your final answers from part *b* in a spreadsheet. Try this different format: display the three car types in rows instead of columns, and then use the other categories as column headings.

3. Conceptual Exercise—The UFO Files

In a recent year, the Gallup organization conducted a poll asking people about their beliefs in UFO's (unidentified flying objects), extraterrestrial life, and other paranormal subjects.

The results out of 100 people are shown in the table below. In this activity, you will review this data and predict from the sample how many people in the United States are likely to hold similar opinions as the people polled.

Question	Yes	No	No Opinion
1. Have you heard or read about UFO's?	87	13	0
2. Have you ever seen anything you thought was a UFO?	12	87	1
3. Do you think that UFO's have ever visited the earth in some form?	45	39	16
4. Do you think there is life of some form on other planets in the universe?	72	19	9
5. In your opinion, does the U. S. government know more about UFO's than they are telling us?	71	19	10

a. From the information in the chart, set up the following ratios. The first one is done for you.

Question Number	Ratio Stated in Words	Ratio in Numbers
1	number of people who have heard or read about UFO's *to* total number of people polled	87 : 100
2	number of people who think they have seen a UFO *to* total number of people polled	
3	number of people who think UFO's have visited Earth *to* total number of people polled	
4	number of people who think there is life somewhere other than Earth *to* total number of people polled	
5	number of people who think the U.S. government withholds information about UFO's *to* total number of people polled	

b. Suppose that you live in a city of 272,000 people. Based on the ratios from part *a*, set up and solve proportions to determine about how many people in your home city would answer "yes" to each of the five questions on the previous page.

Question Number	Set Up the Proportion	Show Work	Number Who Would Answer "Yes"
1			
2			
3			
4			
5			

c. If it were possible to ask everyone in the United States these questions, do you think the number of people who would answer "yes" would be about the same as the ratios from part *a* would indicate? Explain why or why not.

____

____

____

____

____

4. Group Activity—Eating Well

Food	Quantity	Protein (grams)	Calories
Egg white	0.6 ounces (1)	4	16
Skim milk	1 cup (8 oz)	8	86
Pink Salmon	3 ounces	21.6	146
Provolone Cheese	1 ounce	7	100
Red Snapper	3 ounces	19.6	89
Ground beef, extra lean	3 ounces	21	170
Yogurt, low-fat plain	1 cup (8 oz)	13	159
Whole egg	1.6 ounces (1)	7	75
Chicken breast, no skin	3 ounces	26.7	142

a. Using the information in the chart above, calculate the ratio of protein grams to ounces of food and the ratio of protein grams to calories. Record your answers in columns 2 and 4 of the chart below. Next, express your answers in terms of the number of protein grams per one ounce and the number of protein grams per one calorie. Round results to the hundredth's place.

Food	Ratio of Protein Grams to Ounces of Food	Number of Protein Grams per Ounce	Ratio of Protein Grams to Calories	Number of Protein Grams per Calorie
Egg white				
Skim milk				
Pink Salmon				
Provolone Cheese				
Red Snapper				
Ground Beef, extra lean				
Yogurt, low-fat plain				
Whole egg				
Chicken breast, no skin				

b. Which food has the greatest number of protein grams per ounce? Which food has the fewest number of protein grams per ounce?

Greatest: ______________________ Fewest: ________________________

c. Which food has the greatest number of protein grams per calorie? Which food has the fewest number of protein grams per calorie?

Greatest: ______________________ Fewest: ________________________

c. Dr. McGeeney prescribed a low fat diet for Julia that has between 80 and 100 protein grams a day. Determine how many protein grams are in the meals below to determine whether Julia is staying within the recommended diet. Round results to the nearest gram.
Assume that any vegetables, fruits, and bread have very few (if any) protein grams.

Breakfast:

2-egg omelet with 2 oz Provolone cheese
1 glass skim milk (8 ounces)
2 pieces of toast with jam

Total protein grams for breakfast:

Lunch:

Grilled chicken salad with 5 ounces skinned chicken breast
2 bread sticks

Total protein grams for lunch:

Dinner:

8 ounce ground beef, extra lean
Baked potato
2 servings of vegetables

Total protein grams for dinner:

e. Do the meals have the range of total protein grams that the doctor prescribed? If not, what recommendations would you make to Julia?

____

____

____

____

Chapter 6—Percent
Key Concept Activity #6: Free Throw

1. Extension Exercise—Batting Champ

Going into the last day of the major league season, Babe Boggs and Kirby Lockett had identical batting records of 200 hits in 600 at bats. On the last day of the season, Boggs had 7 hits in 8 times at bat, while Lockett had 9 hits in 12 times at bat. Who do you think won the batting title? *Note:* The batting title goes to the hitter with the highest ratio of "hits" to "at bats."

a. **Calculate the identical batting average of Boggs and Lockett just before the last day of the season. The batting average is the ratio (fraction) of "hits" to "at bats." Express this ratio as a fraction in lowest terms, a decimal rounded to the thousandth's place, and a percent.**

$\dfrac{\textit{hits}}{\textit{at bats}} =$

b. **What is the practical meaning of the batting average when expressed as a percent?**

____

____

____

c. **Calculate Boggs' and Lockett's batting average just for the last day of the season. Express the ratio as a fraction in lowest terms, a decimal rounded to the thousandth's place, and a percent. Who had the most hits on the last day? Who had the higher batting average if we just use the last day statistics?**

Last Day Statistics

Boggs: $\dfrac{\textit{hits}}{\textit{at bats}} =$ Lockett: $\dfrac{\textit{hits}}{\textit{at bats}} =$

Most Hits:

Highest Average:

d. **Which player do you think won the batting title? Explain using complete sentences.**

____

____

____

____

e. Calculate the final batting average for each player at the end of the baseball season. Express the ratio as a fraction in lowest terms, a decimal rounded to the thousandth's place, and a percent. Who had the most hits at the end of the season? Who had the higher batting average at the end of the season?

Final Season Statistics

Boggs: $\dfrac{\textit{hits}}{\textit{at bats}} =$ Lockett: $\dfrac{\textit{hits}}{\textit{at bats}} =$

Most Hits:

Highest Average:

f. Who won the batting title? Explain using complete sentences.

____

____

g. Let's return to the last day of the season when Boggs had 7 hits in 8 times at bat, while Lockett had 9 hits in 12 times at bat. The table below shows how Lockett progressed through the last game. Fill in the last three rows, giving his season batting average after each at bat as a ratio, a decimal, and a percent form. Round the decimal to the thousandth's place.

Time At Bat	**1st**	**2nd**	**3rd**	**4th**	**5th**	**6th**	**7th**	**8th**	**9th**	**10th**	**11th**	**12th**
Outcome	Hit	Hit	Hit	Out	Hit	Hit	Hit	Hit	Out	Hit	Out	Hit
Current Average (ratio)												
Current Average (decimal)												
Current Average (percent)												

2. Open Environment Activity —The Fruits of Labor

From 1957 to 1996, the percent of part-time workers in the U.S. increased from 13% to 18%. This trend sparked controversy over whether U.S. businesses were creating too many part-time jobs. Unions and employees didn't like the lower wages and fewer benefits, while employers wanted lower labor costs and the ability to hire only during peak times of the year.

The following table presents national averages of part-time work versus full-time work in 1996.

National Averages for 1996	Part-time	Full-time
Wage per hour	$7.38	$14.16
Receive health insurance benefits	19%	77%

(*Sources*: U.S. Bureau of Labor Statistics and U.S. Labor Dept.)

a. Estimate the ratio (fraction) of full-time wage per hour to part-time wage per hour. Use complete sentences to explain the practical meaning of this ratio.

____

____

____

b. Which of these fractions: $\frac{1}{3}, \frac{3}{4}, \frac{1}{4}, \frac{2}{3}, \frac{1}{5}$ would give you the closest estimate of

- the percent of part-time workers receiving health insurance benefits?
- the percent of full-time workers receiving health insurance benefits?

c. When comparing part-time and full-time workers, which of the following statements gives the most accurate estimate? Circle the best choice and support your answer in words.

- There are 3 times as many full-time workers receiving health insurance benefits as part-time workers.
- There are 4 times as many full-time workers receiving health insurance benefits as part-time workers.
- There are 5 times as many full-time workers receiving health insurance benefits as part-time workers.

____

____

____

Recreate the table below on another piece of paper OR use the directions on pages ix-x of this workbook to create a spreadsheet of this table for parts *d* and *e*.

d. Suppose the average full-time worker's wage remains constant at the 1996 rate, but the average part-time worker automatically receives a 5% raise every year for the foreseeable future. Complete the table below to show what happens to the part-time worker's wage each year for the next five years. Round each raise to the nearest cent.

Year	**Part-time Wage/hour (dollars)**	**Increase in Wage for Next Year (dollars)**	**Next Year's Wage**
1996	7.38	$0.05 \times 7.38 = 0.37$	
1997	7.75		
1998			
1999			
2000			
2001			

e. In what year will the part-time employee's hourly pay rate surpass the full-time employee if the situation in part *d* remains unchanged? Add rows to your table or spreadsheet to show your additional calculations.

f. In 1996, a large mailing company called Mail and More had part-time workers with wages averaging at $9.65 per hour and full-time worker's wages averaging at $19.95 per hour. How would you compare Mail and More's wages with the national averages on the previous page? Consider the ratio of full-time wage per hour to part-time wage per hour as you write your analysis.

____

____

____

g. According to a Boston Globe newspaper article, a survey of area businesses quoted one supermarket manager as saying, "In my produce department, one good full-timer can do as much work (per hour) as any three part-timers." Using this estimate, if a supermarket pays part-time workers an average of $6.00 per hour, then what do you think would be a fair pay rate for full-time workers? Explain your reasoning using complete sentences.

____

____

____

____

3. Conceptual Exercise —Where's the Beef?

A supervisor at a meat packing plant needs to maintain a certain level of quality control for the product due to federal law. Suppose that the supervisor is in charge of marinated beef. In that department, workers place up to 500 pounds of meat into a tumbling machine and add a certain amount of marinade. To maintain quality, the final product cannot have more than a 20% increase in weight due to adding marinade.

a. **An employee named John uses his memory of past experiences to estimate the amount of marinade added. John has just processed 200 pounds of beef, and the final product of marinated meat weighs 234 pounds. If a meat inspector makes a surprise visit, will the inspector give the company a violation? Show all work, and write your answer using complete sentences.**

____

____

____

____

b. **Suppose that 250 pounds of newly arrived beef (called the green product) is placed in a tumbling machine, and some marinade is added before packaging the beef (the final product) for shipment. What is the maximum possible weight of the marinated meat, if it is to stay within federal guidelines? Show all work and write your answer using complete sentences.**

____

____

____

____

c. The meat inspector arrives at the plant and asks the supervisor to explain what method the plant employees use to calculate the 20% rule. This is the process that he describes:

1) Weigh the green product and record the value in pounds.
2) Take the weight in pounds and move the decimal point of this value one place to the left.
3) Double this number to arrive at the maximum increase in weight.
4) The final product is discarded if it weighs more than the green product plus the maximum increase just found in step 3.

Does this procedure correctly meet the 20% rule required by federal law? Support your written answer by showing how the calculations work on 250 pounds of green product meat.

____

____

____

d. The meat inspector wants the supervisor to create a percent change formula that involves using the words *final product* and *green product*. An employee should be able to substitute the weight of the green product and final product into the formula, calculate the percent change, and then compare the value with 20%. Write this percent change formula below.

e. Test your percent change formula from part *d.* Use the formula in the following two situations.

- A plant employee weighs a fresh piece of meat at 230 pounds before mixing it with marinade. After the marinade is added, the weight of the beef is 272 pounds. Would the final product pass inspection under federal law?

- The same employee weighs a second portion of meat at 235 pounds before adding the marinade. After mixing it with marinade the new weight is 284 pounds. Would the final product pass inspection under federal law?

4. Group Activity—A Crowded India

In 1997, India celebrated its 50th anniversary of independence. The country was created from 600 states into a diverse nation with many ethnic groups that speak 18 official languages and practice 7 major religions. India is a study in contrasts when it comes to food, education, and wealth. Although the country is agriculturally self-sufficient, 67% of children under 5 are malnourished. The literacy rate jumped from 17% in 1951 to 52% in 1997, but there is still a wide gap between men and women and between those who live in urban and rural areas.

In May of 1999, the population was approximately 984,103,860 people, and the national currency, the rupee, had the following exchange rate with the U.S. dollar: 1 dollar = 43 rupees. We will use this data in this activity. Work with your group members to answer the following questions.

a. It is estimated that 52% of Indian people live on less than $1 per day. Calculate the number of people in the population who try to survive each day with less than $1 to spend.

b. Suppose a poor man manages to save $1.25 of extra spending money per month. If a basic pair of shoes cost 700 rupees, then how long will it take him to save enough money to buy the shoes?

c. A woman that makes 2,500 rupees per month has the goal of saving enough money in one year to buy a television worth 21,000 rupees. How much money must she save each month in order to reach her goal? How much money does she have left each month to pay bills and live on? What is this amount in U.S. dollars? Round this amount to the nearest cent.

Monthly Savings (rupees):

Monthly (rupees) to live on: Monthly ($) to live on:

d. India's two largest religions are Hindu and Muslim. If Hindus outnumber Muslims 5.7 to 1 and 80% of the country is Hindu, then what percentage is Muslim?

Percent of Muslims in Population:

Number of Hindus in Population:

Number of Muslims in Population:

Use the following table on health and medicine in India to answer the questions below.
Hint: The symbol ≈ means approximately.

Year	Hospitals	Doctors	Doctors/Hospital
1961	3,100	84,000	$\frac{84,000}{3,100} \approx \frac{27}{1}$
1971	3,900	151,000	
1981	6,800	369,000	
1991	11,200	394,000	
1992	13,700	411,000	

e. **Fill in the fourth column to find the ratio of India's doctors per hospital for each year listed. During which year was the number of doctors per hospital at its highest?**

f. **Why is it helpful to calculate the number of doctors per hospital, rather than just relying on the total number of hospitals and doctors, when doing an analysis of India's health care system?**

____

____

____

____

g. **What other data would be useful in completing an analysis on India's health care system?**

____

____

____

____

Chapter 7—Measurement
Key Concept Activity #7: Sea Turtles

1. Extension Exercise—Survival of the Hatchlings

There are about 29,000 - 40,000 adult female Leatherback sea turtles in the world. On average, a female turtle lays eggs in nests every 2 - 3 years. She nests 4 or 5 times in one season, laying about 90 eggs per nest. Approximately 40 percent of the eggs hatch, and 1 in 2500 hatchlings survive to adulthood.

a. Since female Leatherbacks lay eggs on average every 2 to 3 years, we could roughly guess that in one year about one-half to one-third of the females are nesting. Estimate the fewest number of females nesting in one year, assuming that $\frac{1}{3}$ of the lowest estimate, 29,000 adult females, nest in that year. Round your answer to the nearest whole number. Show your work and answers below.

Least number of females nesting in a year: ________________

b. Estimate the greatest number of females nesting in a year, assuming that $\frac{1}{2}$ of the high estimate, 40,000 adult females, nest each year. Show your work and answers below.

Greatest number of females nesting in a year: _____________

c. Use the information from parts *a* and *b* to show the probable range of Leatherback female turtles nesting each year.

From ____________ to _____________ Leatherback female turtles nest each year.

d. Determine a range of how many eggs are expected to hatch from one female Leatherback turtle in one season (if this is her nesting year). Show your work. *Hint:* the number of nests per year for one turtle × the number of eggs in each nest × 40% = number of turtles hatched for one female Leatherback turtle.

From ____________ to _____________ hatchlings by one nesting female turtle each year.

e. Review the information in steps *c* and *d*, and record the range of expected total number of hatchlings in any year. *Hint:* total number of hatchlings = number of female turtles nesting × number of hatchlings from each turtle.

From ____________ to ______________ Leatherback turtle eggs are hatched each year.

f. If the survival rate of hatchlings is 1 : 2500, then what is the range of numbers of each year's hatchlings expected to survive to adulthood? Set up and solve the proportions to answer the question.

From ____________ to ______________ Leatherback hatchlings are expected to survive to adulthood.

g. Mature Leatherbacks measure anywhere from 1.2 to 1.9 meters long and weigh from 200 to 500 kilograms. The heaviest Leatherback ever recorded was 916 kilograms. Convert these metric measurements and weights to the approximate number of units in the U.S. system. Rewrite the first two sentences of this part using U.S. system equivalents. *Hint:* 1 meter ≈ 3.28 feet (≈ means "is approximately equal to"), and 1 kilogram ≈ 2.20 pounds. Round your U.S. units to the nearest tenth for length and to the nearest whole number for weight. Show the unit fractions and set-ups in your conversions.

1.2 meters ≈ ____________________

1.9 meters ≈ ____________________

200 kilograms ≈ __________________

500 kilograms ≈ __________________

916 kilograms ≈ __________________

2. Open Environment Activity—The Panama Canal

You will use the following facts about the Panama Canal in this activity.

Important Measurements for the Panama Canal:

- On the Atlantic side, each of the three **Gatun Locks** is 1000 feet long and 110 feet wide.
- On the Pacific side, the **Pedro Miguel Locks** are about 5/6 mile long total, and the **Miraflores Locks** span a total length of about 1 mile.
- Between the Pedro Miguel Locks and the Miraflores Locks, a ship travels about 1 mile across the man-made **Miraflores Lake**. The locks raise a ship from the Limon Bay channels at sea level to the Gaillard Cut channels that cut through the continental divide.
- **The Limon Bay channel** runs 6 1/2 miles through a mangrove swamp.
- A ship must travel through **Lake Gatun** for 23.5 miles.
- A ship travels 9 miles through the **Gaillard Cut**.

a. On the Pacific side, each of the three Gatun Locks is 1000 feet long. A football field is 100 yards long. How long is each lock in terms of number of football fields? Round results to the nearest tenth. *Hint:* 1 yard = 3 feet

b. Each of the three Gatun Locks on the Atlantic side is 1000 feet long. On the Pacific side, the Pedro Miguel Locks together are 5/6 mile long, and the Miraflores Locks are about 1 mile long. What is the combined length of all the locks in feet and in miles? Round results to the nearest foot and to the nearest tenth of a mile. *Hint:* 1 mile = 5280 feet

Combined length of the locks: ____________ feet = ____________ miles

c. From part *b* and the facts about the Panama Canal, determine the distance through the different parts of the Canal. Record your lengths in miles.

Gatun Lake: __________ miles Limon Bay channel: ________________ miles

Gaillard Cut: __________ miles Miraflores Lake: ________________ miles

Combined length of the locks (from part *b*): ____________ miles

d. From part *c*, determine the total length of the Panama Canal. Express your answer in miles as well as kilometers. To convert from miles to kilometers, use the appropriate unit fraction. Include units when showing your work. Round results to the nearest tenth.
Hint: 1 mile ≈ 1.6 kilometers

Total length of the Panama Canal: __________ miles = __________ kilometers

e. If a ship goes through the Panama Canal in 9 hours, what is its average speed in miles per hour (mph) and in kilometers per hour (kph)? Round your answers to the nearest mile per hour and kilometer per hour.

Hint: $\text{speed (mph)} = \dfrac{\text{number of miles}}{\text{number of hours}}$ $\text{speed (kph)} = \dfrac{\text{number of kilometers}}{\text{number of hours}}$

Average speed of ship through the Canal: __________ mph = __________ kph

f. According to Navy data, a Navy ship can generally travel about 345 miles per day. If a ship traveled from Europe to California around the tip of South America, the trip would be about 5000 miles longer than the route through the Panama Canal. Determine how many fewer days it would take to travel through the Panama Canal instead of routing around South America. Round your result to the nearest day.

g. Use the directions on pages ix-x of this workbook to create a spreadsheet that organizes the length measurements needed to complete this activity. In this spreadsheet, you will show the standard unit equivalents of each length measurement in one column, and the metric equivalents in another column. Include the total Panama Canal length (from part *d*) in miles and kilometers, your answers to part *c* in miles, and your answers to part *c* converted into kilometers.

3. Conceptual Exercise—Converting Doses

The pharmacist has in stock 20 bottles of amoxicillin with 100 capsules in each bottle plus a partially full bottle with 25 capsules. Each capsule contains 250 milligrams of amoxicillin.

Information: 1 kilogram = 1000 grams. 1 gram = 1000 milligrams.

a. How many total milligrams of amoxicillin does the pharmacy have on hand? What is that amount in grams and in kilograms? Round results to the nearest hundredth place.

amoxicillin on hand: ____________ mg = ____________ g = ____________ kg

b. One kilogram equals how many milligrams? *Hint:* Use your answers from part *a* if you need a reference and set up a proportion to answer the question.

1 kg = ________________ mg

c. Doctor Alvarez prescribed 750 milligrams of amoxicillin per day for seven days. How many total milligrams did the doctor prescribe? What is that amount in grams? Round gram results to the nearest hundredth.

amoxicillin prescribed: ____________ mg = ____________ g

d. Since there are 250 mg of amoxicillin in each capsule, how many capsules should the pharmacist put in the bottle to fill the prescription described in part *c*?

________________ capsules

e. How much amoxicillin is in stock after the prescription described in part *c* is filled? Express your answer in milligrams as well as grams.

amoxicillin in stock: ____________ mg = ____________ g

f. If the prescription is followed, how many capsules are to be taken in one day?

____________ capsules per day

g. The pharmacist must make up a label to tell the patient how many capsules to take and when. If the capsules are to be taken in equally spaced intervals over 24 hours, then what instructions should be written on the label? *Hint:* Tell the patient how many capsules to take at which points during the day.

____

____

____

____

____

____

____

____

____

4. Group Activity—Cold Blooded Animals

Warm-blooded humans have a normal body temperature of about 98.6°F, and a 6°F to 7°F change in body temperature can be life threatening. Like all reptiles, Leatherback turtles are cold-blooded, which means that they can change their body temperature to adapt to their surroundings. However, due to an insulating layer of fat, the Leatherback's body temperature may differ considerably from its surroundings. For example, one Leatherback turtle found in 7°C water had a body temperature of 25°C. Work with your group members to answer the following questions.

To convert from degrees Celsius to degrees Fahrenheit, use the formula:

$$F = \frac{9}{5}C + 32$$

a. Using the formula above to convert 7°C and 25°C to degrees Fahrenheit. Show your work below, and round your answers to the nearest degree. Illustrate both temperatures in degrees Celsius and degrees Fahrenheit by drawing a dot at the appropriate points on the scales below. *Hint:* The temperatures corresponding to 7°C and 25°C should be at the same horizontal level on both the Celsius and Fahrenheit scales. If not, check your calculations.

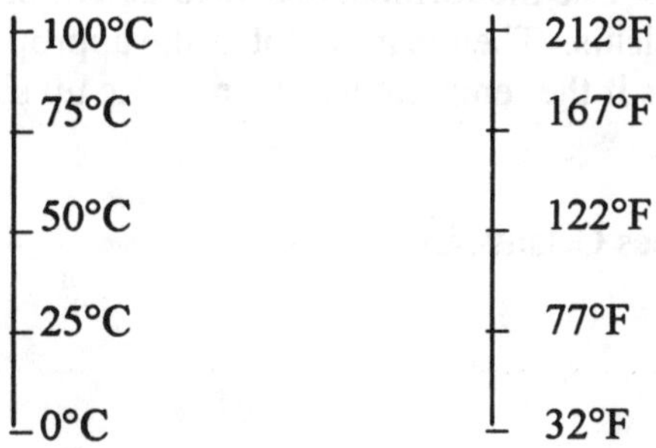

7°C = ____________ °F   25°C = ____________ °F

b. What is the difference between the temperature of 7°C water and the body temperature of the turtle (25°C)? Express this difference in degrees Celsius. Then find the difference between the equivalent temperatures in degrees Fahrenheit. Show your work in the space below.

Difference in degrees Celsius: __________ Difference in degrees Fahrenheit: ___________

c. Write the boiling point (100°C) and freezing point (0°C) in degrees Fahrenheit. Then find the difference in the boiling and freezing points in degrees Celsius and in degrees Fahrenheit.

Boiling Point (°C)	Boiling Point (°F)
Freezing Point (°C)	Freezing Point (°F)

Difference in degrees Celsius: __________ Difference in degrees Fahrenheit: ________

d. Using the information from part *c*, set up and solve a proportion to determine how many degrees Fahrenheit correspond to one degree change on the Celsius scale. Round your results to the nearest tenth of a degree.

e. The normal temperature for people is 98.6°F. Use the formula below to determine the normal temperature for people in degrees Celsius. Then mark a dot at the appropriate points on the scales below. Check your calculations if the temperatures do not line up side by side on the two scales.

To convert from Fahrenheit degrees to degrees Celsius, use:

$$C = \frac{5}{9}(F - 32).$$

98.6°F = ____________ °C

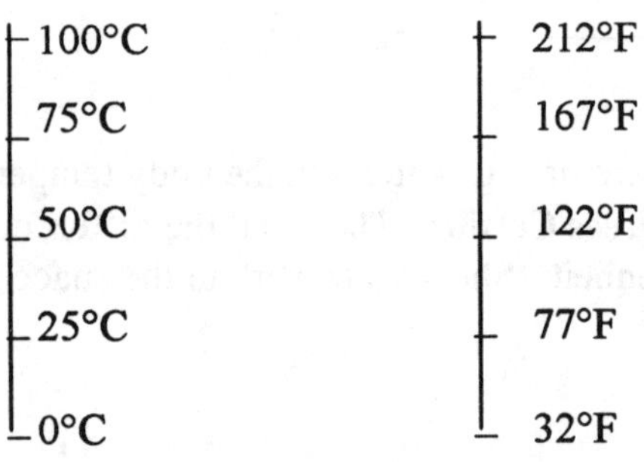

Chapter 8—Geometry
Key Concept Activity #8: Home Run

1. **Extension Exercise —A New Tile Floor**

The rectangular grid below represents tiles of parquet flooring for your new house. Each parquet tile measures one square foot and will be put down in a rectangular room whose floor measures 9 ft. by 12 ft.

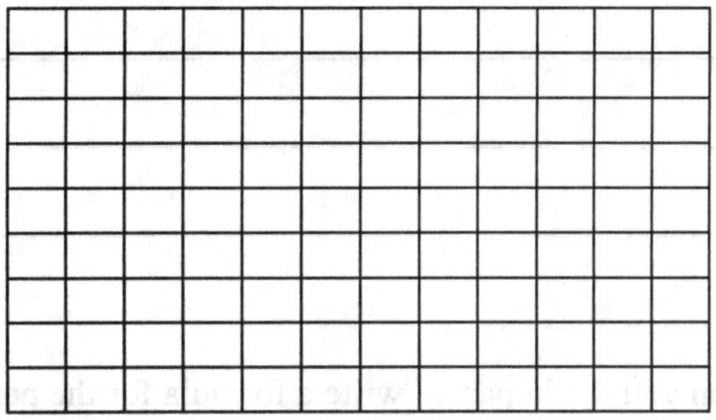

a. **If you walk around the perimeter (all four sides) of the room, how far have you traveled in feet?**

b. **Use complete sentences to explain what you believe is the easiest procedure to calculate the perimeter.**

c. **Draw a picture of the rectangular floor. Label the sides of the rectangle as either *length* or *width*, and then indicate whether the side is 9 ft or 12 ft.**

d. If each tile is one square foot, then how many square feet will cover the entire floor? Show how you can calculate the answer without counting every tile. Explain your work using complete sentences.

___

___

___

___

e. Based on the calculation you did in part *a*, write a formula for the perimeter of a rectangle using the letters *L* for length, *W* for width, and *P* for perimeter. Try to express the formula in two different, yet equivalent, formats.

f. Based on the calculation you did in part *d*, what would be a good formula to use in finding the area of any rectangle? *Hint:* Use the letters *L* for length, *W* for width, and *A* for area.

g. If the floor is 9 ft. by 12 ft., then what would be the units for the area of the floor?

2. Open Environment Activity —Fence Me In

Suppose you just bought a house that sits on a rectangular lot with an area of 7200 square feet. The house is in good shape, but you want to enclose the borders with a fence. You know that the lot is 7200 square feet, and yet you still don't know the actual length or width of the lot.

a. **Use the given data below to complete the table and list some possible dimensions for your 7200 square foot lot. Draw a picture of the situation to help you visualize the problem.**

$$Area = Length \cdot Width$$
$$Perimeter = 2 \cdot (Length + Width)$$

Table with some given lengths:

Area	Length	Width	Perimeter
7200	40		
7200	60		
7200	72		
7200	80		
7200	90		
7200	100		
7200	120		
7200	180		

b. **In the above table, what happens to the width as the length increases? Answer using complete sentences, and explain any patterns you observe.**

___

___

___

___

c. **In the table above, what happens to the perimeter as the length increases? Answer using complete sentences, and explain any patterns you observe.**

___

___

___

___

d. Suppose you want to enclose the entire perimeter (border) of your property with 6 foot high fencing that cost $5.00 per foot of length purchased. The total length of fence you need depends upon the perimeter. If you could choose any length/width pair from the previous table, what dimensions would minimize the cost of fencing in your property? Justify your answer using complete sentences.

____

____

e. Do you think there are dimensions not listed in the table that would decrease the cost of fencing even more than your answer to part *d*? If yes, approximate the length and width. Explain your answer using complete sentences.

____

____

f. Let's investigate by looking at another table that starts with a length of 80 ft and goes up in 1-foot increments to a length of 90 feet. Create a spreadsheet version of this table by using the directions for making spreadsheets on page ix-x of this workbook. Fill in the columns for width, perimeter, and fencing cost using the equations below. Round results to the nearest hundredth.

$Width = \frac{7200}{Length}$ $\quad Perimeter = 2 \cdot (Length + Width) \quad Fencing\ Cost = \$5 \cdot (6 \cdot Perimeter)$

Area	Length	Width	Perimeter	Calculations	Fencing Cost
7200	80				
7200	81				
7200	82				
7200	83				
7200	84				
7200	85				
7200	86				
7200	87				
7200	88				
7200	89				
7200	90				

g. Which perimeter gives the lowest fencing cost? If width values had to be rounded to the nearest foot, which dimensions would result in the lowest fencing cost?

____

____

____

3. Conceptual Exercise—The "Right" Triangle

A long time ago, the Greeks observed that you could construct a 3 by 3 square of tiles, a 4 by 4 square of tiles, and a 5 by 5 square of tiles in such as manner that a right triangle was formed. See the figure below.

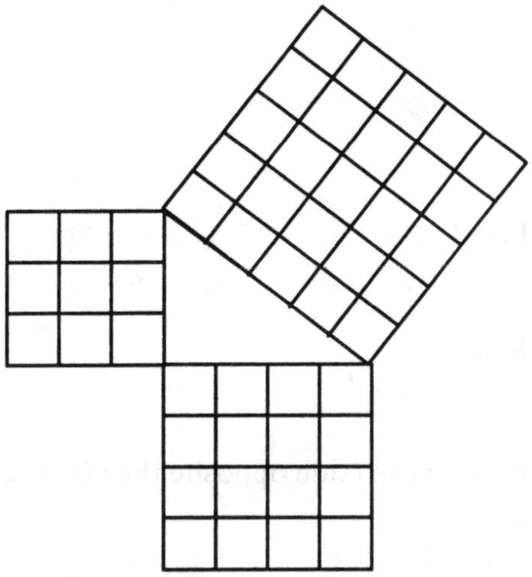

a. What is the area of each square?

3 x 3 square ______________

4 x 4 square ______________

5 x 5 square ______________

b. Using the area for each major square above, write an equality that shows the relationship between the three numbers.

c. What is a simpler way to figure out the areas of each major square? Explain.

____

____

____

d. Try the method that you describe in part *c*. Does the equality from part *b* still hold true if you use the method in part *c*?

____

e. Look back at the right triangle formed by the three squares of tiles and the figure below.

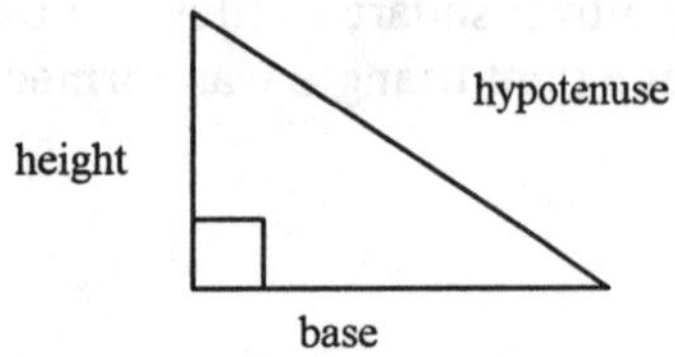

If each tile is one square unit, then

- What is the unit length of the base?

- What is the unit length of the height?

- What is the unit length of the hypotenuse (side opposite the 90° angle)?

f. What is the mathematical relationship between the areas that you found in part *a* and the lengths of the sides in part *e?*

____

____

____

g. Now use the relationship in part *f* to rewrite the equality found in part *b*. *Hint:* Write a statement of equality that generalizes the relationship between the three sides of a right triangle using words and mathematical operations.

4. Group Activity—Composting

To keep your garden soil rich with organic matter, you decide to keep your yard waste in a compost pile. Building a bin using extra wood and wire will promote faster decomposition. A good compost bin is shaped like a cube and has a volume ranging from 27 to 125 cubic feet ($ft.^3$). Work with your group members to answer the following questions.

Materials: paper, ruler

a. A die that you might toss in a board game is an example of a cube. How many numbered faces (sides) does it have?

b. Each face on a cube is an identical square. Represent the compost bin by drawing a wire frame cube that shows all the edges (lines) that would make up the 6 faces. Refer to a picture of such a cube, if necessary.

c. What is the range of length measurements for each identical face (side), if we are to stay within the recommended limits for the volume of the compost bin? *Hint:* $ft \times ft \times ft = ft^3$.

d. Let's make two scaled-down models of the smallest and largest possible bins (with no lid). Use a piece of paper along with scissors and a metric ruler to construct a scaled down model of our cubic bin, with a measure of 1 inch being equivalent to 1 foot of the real bin. To model the smallest recommended bin, use the dimensions, $3\ in \times 3\ in \times 3\ in$, giving a volume of 27 cubic inches (in^3). Follow the steps below.

1.) Draw a square with each of the 4 sides having a length of 9 in.

2.) From each of the 4 corner points (vertices), measure and mark 3 inches in the horizontal direction and 3 inches in the vertical direction.

3.) Use your ruler to draw a 3 inch by 3 inch square in each of the 4 corners.

4.) Use your scissors to cut out each 3 inch by 3 inch square in each of the 4 corners.

5.) Fold up all of the sides to form an open box that will model our compost bin.

Then model the largest possible bin using the dimensions, 5 inches by 5 inches by 5 inches giving a volume of 125 cubic inches.

e. Which of the following two choices would be the most economical for composting? Answer using complete sentences and support your answer with mathematical reasoning.

- 1 compost bin that has a volume of 125 cubic feet.

Or

- 4 compost bins each having a volume of 27 cubic feet.

____

____

____

____

f. How many cubic feet of volume can you construct, if you only have enough material to make a compost bin measuring, $3\frac{1}{2}in \times 3\frac{1}{2}in \times 3\frac{1}{2}in$.

g. Suppose you can carry $\frac{1}{2}$ cubic foot of lawn waste in each full shovel. If you end up with 80 full shovels of lawn waste, will the bin that you constructed in part *f* overflow? Explain.

____

____

____

h. What are the dimensions of a cubic compost bin that has a volume of 64 cubic feet? Explain how you found these dimensions.

____

____

____

Chapter 9—Statistics and Probability
Key Concept Activity #9: The Sneeze Factor

1. Extension Exercise—Winning BigBucks Lottery

Steve and Kate are wondering what 6 numbers to choose for Saturday's lottery game called BigBucks. This is a 6/49 lottery, meaning you play by choosing 6 different numbers between 1 and 49 inclusive and hope that your selection matches the winning combination (randomly picked by the State Lottery Commission).

Steve will use a number obtained by calling 1-900-PSYCHIC to try and win the $2,674,763 jackpot. Kate decides on the following 6-number combination,

1, 2, 3, 4, 5, 6

Her logic is that since any 6-number combination is equally likely to occur, then the above pick has as good a chance of winning as any other combination.

a. Let's start by exploring the six-number combination you must select to attempt winning a 6/49 lottery. How many possible numbers are available when choosing the first number?

b. Once the first number is chosen, how many numbers do you have available when choosing a second number (that must be different from the first number)?

c. How many ways are there to choose the first two numbers? Write your answer as a product of two numbers, and then compute the result.

d. After the first two numbers are chosen, how many numbers are available when choosing a third number (that must be different from the first and second number)?

e. How many ways are there to choose the first three numbers? Write your answer as a product of three numbers, and then compute the result.

f. Knowing that there are $49 \cdot 48 \cdot 47 = 110{,}544$ ways to choose the first three numbers, what do you guess is the number of ways to choose the entire six-number combination needed to win BigBucks? Write your answer as the product of six numbers, and then compute the result.

In the last six questions, we have treated the order of appearance of each number as being important (different). For example, we have counted three number combinations such as 48 7 16 as being different from 7 16 48.

In fact, when playing the lottery, the order in which the numbers appear is not important, so we have over-estimated the number of ways to choose the entire six-number combination necessary to win BigBucks lottery. Let's try to figure out what value must be divided out, so that we are counting only the different combinations without regard to any particular order.

g. Answer the following questions to find how many ordered arrangements (or permutations) of 6 numbers exist.

- Given six distinct numbers, how many ways can you choose the first number? ________
- Having chosen the first number, how many ways can you choose the second number?
- Having chosen the first and second numbers, how many ways can you choose the third number?
- Having chosen the first, second, and third numbers, how many ways can you choose the fourth number?
- Having chosen the first ,second, third, and fourth numbers, how many ways can you choose the fifth number?
- Having chosen the first ,second, third, fourth and fifth numbers, how many ways can you choose the sixth number?
- Altogether, how many ways can you order six numbers?

h. Find the quotient of the product from part *f* divided by the product from part *g*. The result is all the possible six number combinations that have an equally likely chance to occur.

$$\text{Number of possible outcomes} = \frac{\textit{product from part f}}{\textit{product from part g}} =$$

i. If a favorable outcome is selecting the winning number with your ticket, then how many favorable outcomes exist?

Number of favorable outcomes from your ticket:

j. Use the probability ratio to find the chance (probability) of your ticket winning the BigBucks lottery game.

$$\textit{Probability of winning} = \frac{\textit{Number of favorable outcomes}}{\textit{Number of possible outcomes}} =$$

2. Open Environment Activity—What Is Average?

Mr. Smith is ready to give back the midterm exam to his class of 32 students. A student asks him, "How did the class do?" Mr. Smith displays the following chart on the overhead projector.

56	72	74	67	74	92	81	76
82	90	75	98	61	88	65	94
63	58	82	50	77	83	78	86
69	82	61	75	85	78	61	72

a. **Mr. Smith asks you to complete the frequency distribution table given below. You may turn your completed table into a spreadsheet using the Open Environment and the directions for creating a spreadsheet on pages ix-x of this workbook.**

Class Intervals (Exam Scores)	Tally	Class Frequency (Number of Students)
50–59		
60–69		
70–79		
80–89		
90–99		
TOTAL Number of Students		

b. **Next Mr. Smith asks you to construct a histogram from the frequency distribution table in part *b*. *Hint:* The difference between tick marks on the vertical scale should be 1 student. You may want to use the Histogram tool in the Open Environment to create an electronic version of this histogram. To do this, follow the directions on pages ix-x of this workbook.**

Number of Students

Exam Scores

c. Suppose that Mr. Smith uses the following system to convert numerical scores to letter grades:

A: 90-100 *B*: 80-89 *C*: 70-79 *D*: 60-69 *F*: 0-59

- What percent of the students in the class received an *A*?
- What percent of the students in the class received a *B*?
- What percent of the students in the class received a *C*?
- What percent of the class passed (60-100) the midterm?
- What percent of the students failed (0-59) the midterm?

d. Use the information from parts *a* and *c* to create a pie graph (circle graph) of the percentages of A's, B's, C's, D's, and F's in the class. You may draw your graph below or use the directions on pages ix-x of this workbook for creating a pie graph in the Open Environment.

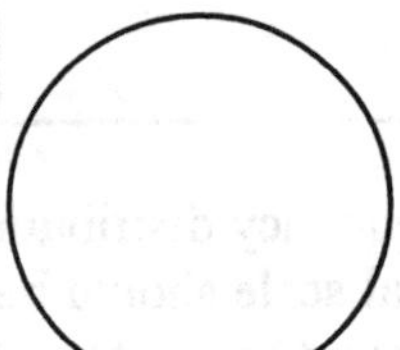

e. For the given set of exam scores,

- find the mean score of the class.
- find the median score of the class.
- find the mode score of the class (there may be more than one).

f. Which measures of central tendency best describe the exam scores of the class? Explain using complete sentences.

____

____

____

3. Conceptual Exercise—Lightning Strikes

Suppose we want to compare the probability of winning a 6/49 lottery, with the probability of a randomly selected American being hit by lightning, over a one-year time period. The resident population of the United States is approximately 273,000,000. According to the National Center for Health Statistics, lightning kills between 200 to 300 people in the U.S. each year.

a. Let's round the U.S. population to 270,000,000 and use 270 as the amount of deaths caused by lightning per year. What is the probability that you will be hit by lightning this year? State the probability as a fraction, a decimal rounded to the hundredth's place, and a percent. Use complete sentences to explain your answer.

Fraction:

Decimal:

Percent:

____

____

____

b. Do you think that every resident of the United States has the exact same probability of being hit by lightning? Use complete sentences to explain your answer.

____

____

____

____

c. **If you round the chances of winning the lottery to 1 in 14 million, then how many more times likely are you to be hit by lightning than to win the lottery? Show all work and explain your answer using complete sentences.**

____

____

____

d. **Use complete sentences to give your opinion on the following statement.**

The lottery is the state's way to get an additional tax on the innumerate citizen.

Hint: The word "innumerate" means "weak in math."

____

____

____

____

____

4. Group Activity—Laundry Blues

Tom woke up on a rainy Monday morning with nothing to wear for class. He forgot to do his laundry yesterday. It turns out that his roommate John actually did some of his laundry after all. There are 4 clean shirts and 3 clean pairs of pants hanging in the closet. Tom thinks that he doesn't have enough choices to mix and match into a suitable outfit to wear, but John says that there are 12 possible outfits to choose from, and at least one of these outfits must match!

a. Let's define the 4 shirts as S#1, S#2, S#3, and S#4, and the 3 pairs of pants as P#1, P#2, and P#3. Use a tree diagram to list the 12 possible outfits that Tom can choose from.

b. Suppose there are 8 possible combinations that Tom would consider suitable. If he walks into the dark closet and randomly chooses one pair of pants and one shirt without being able to distinguish colors or styles, then what is the probability that Tom has picked an outfit that he considers to be suitable? Show all work, and state your answer as a fraction, a decimal rounded to the hundredth's place, and a percent.

Fraction: Decimal: Percent:

c. What is the probability that in the dark closet Tom has randomly picked out an outfit that is not suitable? Show all work, and state your answer as a fraction, a decimal rounded to the hundredth's place, and a percent.

Fraction: Decimal: Percent:

d. Assume that S#1 is Tom's favorite shirt. What is the probability of randomly picking this shirt in the dark closet? Show all work, and state your answer as a fraction, a decimal, and a percent.

Fraction: Decimal: Percent:

e. Suppose there is one shirt with two buttons missing and 1 pair of pants with a broken zipper, what is the probability that Tom has picked:

- both these items?

  Fraction: Decimal (to nearest hundredth): Percent:

- one of them?

  Fraction: Decimal (to nearest hundredth): Percent:

f. Tom finally chooses an outfit. While getting dressed, he listens to the radio and hears that there is a 75% chance of rain. Does this mean that rain will fall from the sky 75% of the time that day? Support your answer using complete sentences.

____

____

____

____

____

____

Chapter 10—Signed Numbers
Key Concept Activity #10: Fire and Ice

1. Extension Exercise—Temperature Scales

If C is the temperature in degrees Celsius, and F is the temperature in degrees Fahrenheit, then we can represent the relationship between temperature scales with the following formula.

$$C = \frac{5}{9} \cdot (F - 32)$$

a. Convert the Fahrenheit temperatures of 32°F, 23°F, and 14°F to the Celsius scale. Show the two-step process needed to arrive at your answer.

$\frac{5}{9} \cdot (32 - 32) =$ $\frac{5}{9} \cdot (23 - 32) =$ $\frac{5}{9} \cdot (14 - 32) =$

b. Use complete sentences to describe the two steps you performed in completing part *a*.

Step 1: ____

____

Step 2: ____

____

c. Let's reverse the above procedure by first describing the opposite of step 2 and then the opposite of step 1. This new two-step procedure should allow you to convert any Celsius temperature to Fahrenheit temperature.

Reverse of Step 2: ____

____

Reverse of Step 1: ____

____

d. Show how to convert the Celsius temperatures of 0°C, –5°C, and –10°C to Fahrenheit temperatures using the reverse procedure developed in part *c*.

e. Letting C be the temperature in degrees Celsius and F be the temperature in degrees Fahrenheit, suggest a formula for converting from degrees Celsius to degrees Fahrenheit.

f. Temperatures in space are measured in degrees Kelvin. Find the relationship between the Kelvin scale and the Celsius scale by observing the last two columns.

Temperature of	Kelvin	Celsius
Earth's Sun	6000	5727
Water Boiling	373	100
Water Freezing	273	0
Absolute Zero	0	–273

- What operation must you perform to convert a Kelvin temperature to a Celsius temperature?

- What operation must you perform to convert a Celsius temperature to a Kelvin temperature?

g. If K is a temperature in degrees Kelvin and C is a temperature in degrees Celsius, then suggest a formula for converting degrees Kelvin to degrees Celsius and a formula for converting degrees Celsius to degrees Kelvin.

2. Open Environment Activity —Balancing Your Account

You just received your bank statement, which shows an ending balance of $25 in your checking account. Where did all the money go? You decide to compare last month's deposits and withdrawals. A good way to do this is to take the balance from the start of last month. Then add a positive number for deposits, and add a negative number for withdrawals. The following table shows this process, and the first calculation is completed for you.

	Transaction	Withdrawal (–) ($)	Deposit (+)	Calculation	Balance ($)
	Start of Month				+625
	Rent	-550		+625 + (-550)	+75
	Car Loan	-250			
	Hair Cut	-25			
	Paycheck		+600		
	Auto Repair	-575			
	Electric Bill	-75			
	Telephone Bill	-110			
	Cable	-25			
	Paycheck		+600		
	Groceries	-190			
TOTALS →	End of Month				+25

a. Finish filling in the table by showing the calculation needed to obtain the balance after each transaction. Place the result of each calculation in the balance column as either a positive or negative number. Complete all calculations by hand, and then go back to check your work with a calculator or spreadsheet on the Open Environment.

b. Observe all the calculations where you added a positive number and a negative number.

- If you temporarily ignore the signs of the numbers, then what operation did you actually perform on the two numbers?
- After performing the operation, what about the two numbers determined whether the balance was positive or negative?
- In your own words write a general procedure for adding positive and negative numbers.

____

____

____

____

____

c. Now observe the calculations where you added two numbers with the same sign (either both are positive or both are negative).

- If you temporarily ignore the signs of the numbers, then what operation did you actually perform on the two numbers?
- After performing the operation, what about the two numbers determined whether the balance was positive or negative?
- In your own words write a general procedure for adding two numbers with the same sign.

____

____

____

____

____

d. Return to the table and add up all the withdrawals in the second column. Record the sum below and in the table (*end of the month* row, *withdrawal* column).

e. Return to the table and add up all the deposits in the third column. Record the sum below and in the table (*end of the month* row, *deposit* column).

f. Analyze the results of parts *d* and *e*. Then explain what is happening to your cash flow over this one-month period.

____

____

g. How did you manage to end up with +25 dollars when more money was going out than coming in?

____

____

h. Assuming you have no auto repairs or hair cuts next month, estimate your ending balance for next month. Show all your work.

3. Conceptual Exercise —Rent-a-Calculator

Suppose you decide to start a business renting graphing calculators to other students at your college. Before you can begin your business, you need to purchase some graphing calculators. You have enough money saved to buy 50 reconditioned calculators from a wholesaler for $45 per calculator. On campus, you advertise that any student can rent a calculator for one semester at a fee of $15 per semester, but you must return it in good condition. You plan on running this business over a 2-year period that will cover 6 semesters (fall, spring, and summer of each year).

a. What is the initial *cost* to start your business? Show your work.

b. If you manage to rent all 50 calculators for 2 consecutive semesters, what will be your *revenue* at the end of this time period? Revenue is the money coming in from the rentals. Show your work.

c. Use your results from parts *a* and *b* to calculate the *profit* after two semesters. Show your work.

$$Profit = Revenue - Cost$$

d. If the *profit* is negative, then explain the status of the business in terms of revenue and cost. Use complete sentences in your explanation.

____

____

____

____

e. How many calculators must you rent during your third semester of business to break even? Show all work.

f. By the end of the third semester, your business does break even. Now you set the goal of making \$3000 profit over the next 3 semesters. To accomplish your goal, what is the new minimum rental price that must be charged for the calculators?

g. What type of risk are you taking by charging a price greater than \$15 dollars? Explain using complete sentences.

____

____

____

____

h. Suppose the increase in price causes a decrease in demand such that you are only able to rent 40 calculators in each of the last three semesters of running your business. How much are you short of your \$3000 goal if you use the rental price from part *f*?

4. **Group Activity—Elevator Talk**

Suppose you work in a building that has three floors above ground level and two floors below ground level. Imagine that the figure below represents an elevator that takes people up and down to different floors. Assume that the distance between successive floors is 10 feet.
Work with your group members to answer the following questions.

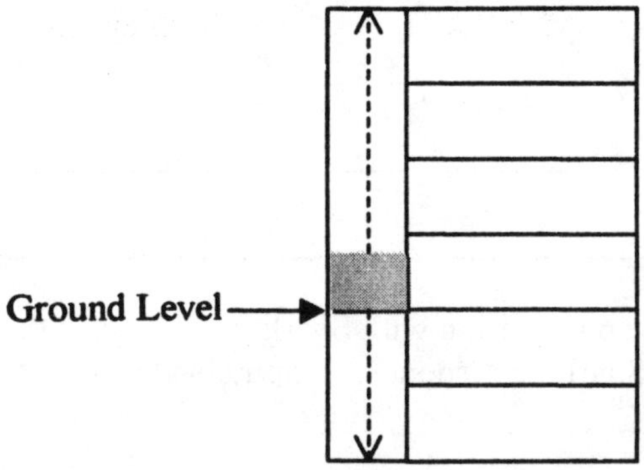

a. Label the ground level floor as 0, and then label the other floors using a positive or negative integer.

b. Sharlitha enters the elevator at ground level, travels up to floor 1 to drop off a letter, and then goes down to floor −2 to pick up some supplies. To the right of the figure above, draw and label two arrows pointing up or down to visualize how positive and negative numbers can represent her vertical movement on the elevator.

c. Use addition of positive and negative integers to show the calculation that represents Sharlitha's elevator trip and the solution that represents the floor she ends up on (floor −2).

d. Let 10 represent the number of feet Sharlitha traveled up and −30 the number of feet she traveled down. Use subtraction to represent the total distance that Sharlitha traveled between floors.

e. After picking up supplies on floor –2, Sharlitha gets back on the elevator with two other people, Terry and Marco. Marco says to Terry, "I do not understand how these floors are numbered. If zero means *nothing*, then it's impossible for anything to be less than nothing." State whether you agree or disagree with Marco, and use complete sentences to support your answer.

____

____

____

____

f. Terry gets out on floor –1. Then Marco starts up a conversation with Sharlitha saying, "Oh, I have the money I borrowed from you yesterday." Sharlitha responds, "Happy birthday! Consider that money as my treat for lunch."

If b represents the amount of money that Marco borrowed from Sharlitha, then how would you interpret the following rule? $-(-b)=b \quad (b>0)$
Hint: The great mathematician Leonard Euler once said, …*to cancel a debt signifies the same as giving a gift.*

____

____

____

g. Marco and Sharlitha leave the elevator at floor 0. Before going back to work, Marco describes his five-month battle trying to lose weight. "In the first month of dieting, I cheated all the time and gained 10 pounds. During the second month, I lost 5 pounds. Then, in the third month I lost 7 pounds. During the fourth month I lost 3 pounds, and now I have gained 4 pounds during my fifth month." Marco refers to his weight journal below:

Month	January	February	March	April	May
Weight Change	+10	–5	–7	–3	+4

- Show the calculation needed to find the total change in weight over the 5-month period.

- How much did Marco gain or lose? Express your answer as a signed number.

Chapter 11—Introduction to Algebra
Key Concept Activity #11: A Daily Dose

1. Extension Exercise—Mystery Dose

It is difficult to calculate the exact surface area of a human body since it is not flat or shaped like a box. However, health professionals have come up with a formula to approximate the number of square meters it would take to cover a human body. All you need to determine a person's body surface area or BSA is that person's height and weight. Once calculated, a person's BSA can be used to determine the proper individual dose for some medications.

In the chart below, we have data on 2 people. In each case, either the BSA or the proper dose is unknown and has been replaced by a variable, x.

Person	Body Surface Area (square meters)	Recommended Rule for Finding the Proper Dose	Proper Dose (milligrams)
Linda	1.5	3.7 mg. per BSA of 1 sq. meter	x
Jim	x	9.3 mg. per BSA of 1 sq. meter	23.3

a. For Linda, write a general formula to solve for x. Use words and the variable x in your formula. Label what x represents by completing the "Let x =" statement below. Then substitute the known values into your formula, and solve for x. Round results to the nearest tenth of a milligram. Show your work below.

Let x = ____

General formula for Linda: x = ________________________

b. For Jim, write a general formula to solve for x. Use words and the variable x in your formula. Label what x represents by completing the "Let x =" statement below. Then substitute the known values into your formula, and solve for x. Round results to the nearest tenth. Show your work below.

Let x = ____

General formula for Jim: x = ________________________

c. For the next five days, Linda's dose needs to slowly increase. Use Linda's BSA from the chart on the previous page and the table below to find the proper dose (in milligrams) for Linda on each day. Round results to the nearest tenth of a milligram. Show calculations and record answers in the table below.

Milligrams per 1 square meter of BSA	**Calculations for Linda**	**Amount of Dose (milligrams)**
First day: 3.7		
Second day: 5.5		
Third day: 7.4		
Fourth day: 9.3		
Fifth day: 11.1		

d. Jim's dose each day needs to slowly decrease for the next five days. Use Jim's BSA from your answer in part *b* and the table below to find the proper dose (in milligrams) for Jim on each day. Round results to the nearest tenth of a milligram. Show calculations and record answers in the table below.

Milligrams per 1 square meter of BSA	**Calculations for Jim**	**Amount of Dose (milligrams)**
First day: 9.3		
Second day: 8.6		
Third day: 7.5		
Fourth day: 5.4		
Fifth day: 3.5		

2. Open Environment Activity—Community Fund-Raiser

Your community is hosting a fund-raiser for charity. Decorations, pamphlets, and other miscellaneous items have been donated. Volunteers have planned the event and will prepare the food, so there are no labor costs for the event. It will cost about $5.25 per person to buy the food, and $320 to rent a hall for the evening.

a. Do the necessary calculations and fill in the following table to show how much the event will cost depending upon the number of people attending in the first column.

No. of People Attending	Cost of Room Rental ($)	Cost of Food, $5.25 per person ($)	Total Cost ($)
50			
100			
150			
200			
250			
300			
350			

b. Write an expression that represents the cost of food if *n* people attend the fund-raiser.

cost of food for *n* people: ____________

c. Write an expression that represents the total cost if *n* people attend the fund-raiser.

total cost for *n* people: ____________

d. If C represents total cost and n represents the number of people attending the fund-raiser, write an equation that relates C and n. *Hint:* Use your expression from part *c*.

equation: __________________________

e. If people are charged $15 each for the dinner, calculate the total revenue for the dinner. Then copy the total cost from the table in part *a* into the 3rd column, and calculate the profit, which is revenue minus total cost. Round dollar amounts to the nearest cent. Use the directions on pages ix-x of this workbook to turn this table into a spreadsheet on the Open Environment.

No. of People Attending	Revenue = $15 per person ($)	Total Cost ($) (from part a)	Profit = Revenue – Cost ($)
50			
100			
150			
200			
250			
300			
350			

f. If R represents revenue and n represents the number of people attending the fund-raiser, then write an equation for R in terms of n.

$R =$ ______________________________

g. Record your equations from parts *d* and *f* in the spaces provided below.

$R =$ ______________ $C =$ ______________

h. The profit of the fund-raiser will be the revenue minus the cost. Use the formulas in part *g* to write an expression for profit P. Then simplify the expression by removing parentheses and combining like terms.

$P =$ _______________ – (_______________) = _______________

Revenue Cost Simplified

i. Use the equation for profit in part *h* to determine how many people would have to attend the event to make a profit P of $800. Round your answer to the nearest whole person.

3. Conceptual Exercise—Money in Trees

When determining the dollar value of a tree, you must estimate the size of the tree. To do this, you need to calculate the tree's diameter breast height (dbh), which is the diameter of the tree trunk at 4 ½ feet above the ground (breast-high on a man of average height).

a. **There is no real monetary value of a tree until it obtains a diameter breast height (dbh) of about 14 inches. Use the formula, $C = \pi d$, to find the minimum circumference C (at breast height) for a tree to begin to have real monetary value. Round your answer to the nearest whole inch.** ***Hint:*** $d = 14$, $\pi \approx 3.14$

Minimum circumference: C = ____________ inches

b. **For each circumference given below, find the diameter breast height of the tree. First, substitute the given value for C into the equation, $C = \pi d$. Then solve the equation for d. Show your substitution and work in the second column, and record your answer for d in the last column. Round d to the nearest inch.** ***Hint:*** $\pi \approx 3.14$

Circumference at Breast Height (inches)	Equation, Show work	What is d? (inches)
47		
53		
66		
88		

c. **If you know the circumference of a tree, how would you rewrite the equation $C = \pi d$ so that it begins as $d = ?$ *Hint:* To check your equation, use it to calculate the diameter breast height d for each circumference in part b and see if you get the same answer.**

$d =$ ____________________

d. **If a tree's diameter decreases 1.5 inches every 12 feet up the tree (on average), then how much will the tree decrease in diameter for each 1 foot increase in height? Round your results to the nearest thousandth's place. *Hint:* You can set up and solve a proportion to figure this out.**

Decrease in diameter per foot = __________ inches

e. **Using the answer from part *d*, explain how you would find the decrease in diameter of a tree at 20 feet.**

____

____

____

f. **Generalize the process that you used in part *e* and write a formula for the decrease in the diameter over *x* feet.**

Decrease in diameter = ____

4. Group Activity—Springs in the Ozarks

Mark, a geologist, recently traveled to the Ozarks in Missouri and Arkansas. He is fascinated by the constant flow of underground springs that feed the many clear, deep pools in the area, and thus, he gathered the following information about 2 of these underground springs: Round Springs and Big Springs, both in Southern Missouri.

Name of Spring	Average Amount of Water Flow Per Day (millions of gallons)	Maximum Water Flow Ever Measured Per Day (millions of gallons)
Big Spring	276	800
Round Spring	26	300

The formula used to calculate the flow rate of a spring is $A = r \bullet t$, where A is the amount of water flow, r is the rate of water flow, and t is the measured amount of time. In the table above, t is equal to one day. When using this formula, the units must match.

Work with your group members to answer the following questions.

a. Solve the formula above for r.

b. Rates are measured in such units as miles per hour, feet per second, gallons per minute, and kilograms per hour. At the springs, what unit is used to measure the flow rate of the water?

c. What is the average flow rate r for Big Springs? Use the data in the 2nd column of the table and the formula for r from part *a*.

Average flow rate, Big Springs: r = ______________ millions of gallons of water per day

d. What is the average flow rate r for Round Springs? Use the data in the 2nd column of the table and the formula for r from part *a*.

Average flow rate, Round Springs: r = ____________ millions of gallons of water per day

e. What is the maximum flow rate *r* for Big Springs? Use the data in the 3rd column of the table and the formula for *r* from part *a*.

Maximum flow rate, Big Springs: r = ___________ millions of gallons of water per day

f. What is the maximum flow rate *r* for Round Springs ? Use the data in the 3rd column of the table and the formula for *r* from part *a*.

Maximum flow rate, Round Springs: r = _________ millions of gallons of water per day

g. Calculate the estimated amount of water to flow from each of the springs in an average year. Give your answers in millions of gallons per year. (1 year = 365 days)

Amount of water through Big Springs: ____________________
millions of gallons per year

Amount of water through Round Springs: ____________________
millions of gallons per year

h. About how many days would it take for one billion gallons of water to flow through each of the springs? Round your answers to the nearest whole day. *Hint:* Convert one billion into millions of gallons. Then substitute the result into the equation for *A*, and solve for *t*.

Big Springs will have one billion gallons of water flow through it in _______________ days.

Round Springs will have one billion gallons of water flow through it in _______________ days.

Applications Index

Applications Index (continued)